NUCLEAR PHYSICS

Other titles in the Project

Physics   Robert Hutchings
Telecommunications   John Allen
Medical Physics   Martin Hollins
Energy   David Sang and Robert Hutchings

Biology   Martin Rowland
Applied Genetics   Geoff Hayward
Applied Ecology   Geoff Hayward
Micro-organisms and Biotechnology   Jane Taylor

Biochemistry and Molecular Biology   Moira Sheehan
Chemistry   Ken Gadd and Steve Gurr

UNIVERSITY OF BATH • SCIENCE 16-19

Project Director: J. J. Thompson, CBE

NUCLEAR PHYSICS

DAVID SANG

Nelson

Thomas Nelson and Sons Ltd
Nelson House   Mayfield Road
Walton-on-Thames   Surrey
KT12 5PL   UK

51 York Place
Edinburgh
EH1 3JD   UK

Thomas Nelson (Hong Kong) Ltd
Toppan Building 10/F
22A Westlands Road
Quarry Bay   Hong Kong

Thomas Nelson Australia
102 Dodds Street
South Melbourne
Victoria 3205   Australia

Nelson Canada
1120 Birchmount Road
Scarborough   Ontario
M1K 5G4   Canada

First published by Macmillan Education Ltd 1990
ISBN 0-333-46658-6

This edition published by Thomas Nelson and Sons Ltd 1992

ISBN 0-17-448208-6
NPN 9 8 7 6 5 4 3

Printed in Hong Kong.

Contents

Theme 3:   **Nuclear technology**

The Project: an introduction

The **University of Bath Science 16–19 Project**, grew out of a reappraisal of how far sixth form science had travelled during a period of unprecedented curriculum reform and an attempt to evaluate future development. Changes were occurring both within the constitution of 16–19 syllabuses themselves and as a result of external pressures from 16+ and below: syllabus redefinition (starting with the common cores), the introduction of AS-level and its academic recognition, the originally optimistic outcome to the Higginson enquiry; new emphasis on skills and processes, and the balance of continuous and final assessment at GCSE level.

This activity offered fertile ground for the School of Education at the University of Bath to join forces with a team of science teachers, drawn and a major publisher from a wide spectrum of educational experience, to create a flexible curriculum model and then develop resources to fit it. This group addressed the task of satisfying these requirements:

- the new syllabus and examination demands of A- and AS-level courses;
- the provision of materials suitable for both the core and options parts of syllabuses;
- the striking of an appropriate balance of opportunities for students to acquire knowledge and understanding, develop skills and concepts, and to appreciate the applications and implications of science;
- the encouragement of a degree of independent learning through highly interactive texts;
- the satisfaction of the needs of a wide ability range of students at this level.

Some of these objectives were easier to achieve than others. Relationships to still evolving syllabuses demand the most rigorous analysis and a sense of vision – and optimism – regarding their eventual destination. Original assumptions about AS-level, for example, as a distinct though complementary sibling to A-level, needed to be revised.

The Project, though, always regarded itself as more than a provider of materials, important as this is, and concerned itself equally with the process of provision – how material can best be written and shaped to meet the requirements of the educational market-place. This aim found expression in two principal forms: the idea of second-ment at the University and the extensive trialling of early material in schools and colleges.

Most authors enjoyed a period of secondment from teaching, which not only allowed them to reflect and write more strategically (and, particularly so, in a supportive academic environment) but, equally, to engage with each other in wrestling with the issues in question.

The Project saw in the trialling a crucial test for the acceptance of its ideas and their execution. Over one hundred institutions and one thousand students participated, and responses were invited from teachers and pupils alike. The reactions generally confirmed the soundness of the model and allowed for more scrupulous textual housekeeping, as details of confusion, ambiguity or plain misunderstanding were revised and reordered.

The test of all teaching must be in the quality of the learning, and the proof of these resources will be in the understanding and ease of accessibility which they generate. The Project, ultimately, is both a collection of materials and a message of faith in the science curriculum of the future.

J.J. Thompson
January 1990

How to use this book

Nuclear physics is a branch of physics which has grown dramatically in the twentieth century. It has grown in two ways. Firstly, experiments and the theoretical interpretation of them have greatly extended our knowledge of the existence of the nucleus and the forces which hold it together. Secondly, our ability to handle radioactive materials has resulted in major new technologies, for good and, sadly, for ill. Nuclear physics has been successful in explaining many things. It can explain the life history of the stars, and the abundances of different elements in the universe. And it may lead us to vast untapped stores of energy for the future.

Nuclear physics, and in particular nuclear technology, has captured the popular imagination. At times, people have been excited by the idea of splitting the atom; at other times, we have been gripped by fear of the consequences of the power over the basic forces of nature which we now have at our disposal. All technologies have this dual nature: we love them and we hate them. I hope that, in learning more about the science behind this technology, you will be able to apply your knowledge to understanding the problems associated with nuclear technology, and make an informed, critical evaluation of its worth.

Nuclear physics is part of all A-level physics syllabuses and this book covers all the requirements of courses at this level. In particular the book has been written for A- and AS-level courses which have options in nuclear physics. I have assumed some knowledge and understanding of radioactivity and the structure of the atom, in particular that radioactive materials gradually decay and that the rate at which they decay may be described by their half-life.

The book is divided into three themes. The first covers our knowledge of the structure of the nucleus, the second deals with the processes of radioactive decay and the final theme, called Nuclear Technology, looks at the uses made of our knowledge of nuclear physics. There are a number of activities throughout the text for you to carry out in addition to reading, which on its own is too passive to promote effective thinking and learning. Questions in the text have two purposes: to help you to check your knowledge and understanding as you proceed, and to encourage you to think ahead, to work out the next steps in the argument. The assignments will help you to practise using your knowledge, or to carry out some simple experimental investigations. At the end of each theme are some examination questions, to help you to assess whether you have reached an appropriate standard in your studying.

I have assumed that you have someone to guide your learning, and to help you out if you get really stuck. I have tried to write this book so that it will support your learning, but if in doubt, ask your teacher.

For most people, it's easiest to study in collaboration with a partner. You can set each other targets, help each other out over the difficult bits, try out the investigations together, and generally make your learning more enjoyable.

Nuclear physics is a mathematical subject. You will need a calculator which has scientific notation, and you will need to know how to use it! Most of the calculations involved are simply adding, subtracting, multiplying and dividing, but you will need to be able to handle very large and very small numbers to a high degree of precision.

Finally, you will need persistence. Nuclear physics has some tricky concepts, which you may find difficult to grasp at first, so persevere. I hope that, by following the text, answering the questions and carrying out the other activities, you will be able to grasp these ideas, which have been among the most influential scientific ideas of the twentieth century.

Learning objectives

These are given at the beginning of each chapter and they outline what you should gain from the chapter. They are statements of attainment and often link closely to statements in a course syllabus. Learning objectives can help you make notes for revision, especially if used in conjunction with the summaries at the end of the chapter, as well as for checking progress.

Questions

In-text questions occur at points when you should consolidate what you have just learned, or prepare for what is to follow by thinking along the lines required by the question. Some questions can, therefore, be answered from the material covered in the previous section, others may require additional thought or information. Answers to numerical questions are at the end of the book.

Examination questions

At the end of each theme is a group of examination questions relating to the topics covered in the theme. These can be used to help consolidate understanding of the theme or for revision at the end of the course.

Investigations

There are not many nuclear physics experiments which you can carry out safely in a school or college laboratory. I have included a few, some of which are models which can help you to build up a mental picture of the subject.

I have not given full details of experimental methods, except where these are necessary for reasons of safety. You should be able to decide for yourself what quantities need to be measured, and how to measure them.

Assignments

Where you are asked to think about a particular idea, or to develop an idea further, you will find text and questions presented together as an assignment. Sometimes these will require you to refer to other resources, some of which are suggested in Appendix A.

Summaries

Each chapter ends with a brief summary of its content. These summaries, together with the learning objectives, should give you a clear overview of the subject, and allow you to check your own progress.

Chapter 10

In the last chapter, you will find four sections which are intended help show you how the ideas of nuclear physics can help you to understand some of the present and future problems of physics.

Other resources

I have assumed that you will have access to various useful books and software; in particular, you will need to have a data book which lists the properties of different radioactive nuclides, and perhaps a chart of radionuclides. I have listed some suitable books and software in Appendix A.

Appendix B lists some useful addresses, and Appendix C contains values of some physical constants which are used very frequently. You will need to refer to Appendix C frequently; it will help you if you make a photocopy of this section and use it as a bookmark.

Acknowledgements

The author and publishers wish to thank the following who have kindly given permission for the use of copyright material:

The Associated Examining Board, Joint Matriculation Board, Oxford and Cambridge Schools Examination Board and University of London School Examinations Board for questions from the past examination papers; Cambridge University Press for material from *The Quantum Universe* by Tony Hay and Patrick Walters, 1987.

The author and publishers wish to acknowledge, with thanks, the following photographic sources:

American Institute of Physics, Niels Bohr Library *pp 21, 31 left, 129 top;* Associated Press *p 87 middle left, 99, 132 bottom;* Australia News and Information Bureau *p 107;* British Antarctic Survey *p 88 bottom left;* British Nuclear Fuels Ltd *p 108 top and bottom, 124 top;* Brookhaven National Laboratory *p 74 top;* CERN *p 130 right, top and bottom left;* Greenpeace *p 124 bottom left;* Griffin & George *p 85 bottom;* Philip Harris *p 85 top;* Michael Holford *p 88 bottom right;* Isotron *p 88 top right;* JET *p 122;* Kobal Collection *p 131;* National Radiological Protection Board *p 84;* David Neal *pp 10,12 top and bottom, 80, 86 top, 96;* Royal Observatory, Edinburgh *p 128;* Royal Society *p 132 top;* Sabre International Products Ltd *p 88 middle left;* David Sang *p 31 right;* reproduced by permission of the Trustees of the Science Museum *pp 63 both photographs, 82;* Science Museum and UKAEA *p 105;* Science Photo Library *pp 1 bottom, 79 top, 86 bottom right, 88 middle right, 129 bottom, 130 top left;* Soviet TV *p 124 bottom right;* TASS *p 125;* Topham Picture Library *p 45;* United Kingdom Atomic Energy Authority *pp 86 bottom left, 87 top left and right, middle and bottom right, 88 top left, 106;* University of Cambridge, Cavendish Laboratory *pp 1, 2, 5, 74 bottom;* University of Manchester *p 6;* West Air Photography *p 87 bottom left;* Andrew Wiard (Report) *p 79 bottom;* Yorkshire Post Newspapers, courtesy of Grenville Needham *p 126.*

Every effort has been made to trace all the copyright holders, but if any have been inadvertently overlooked the publishers will be pleased to make the necessary arrangement at the first opportunity.

Theme 1

PROBING THE NUCLEUS

Nuclear physics could be said to have started with Becquerel's discovery of radioactivity in 1896. By 1911 Rutherford had established the existence of the nucleus. In 1935 Yukawa presented an explanation of the forces which hold the nucleus together. Within four decades the foundations had been laid for our understanding of the nuclear atom.

Subatomic particles, and the particles of which they themselves are made, are the object of continuing research efforts; often this is very expensive, requiring huge machines operated by large teams of scientists and engineers. The role of each individual in a large project can be difficult to understand and the public is not always happy to foot the bill. It is chastening to see the human scale of the laboratory in which Rutherford and his collaborators worked in Cambridge in the 1920s.

You will already be familiar with the idea that every atom has a central core or nucleus. In these two chapters we will look at the way in which this picture of the atom was established, and at other ways in which we can probe the structure of the nucleus. Then we will look at some of what is known about the forces which hold the nucleus together. This will provide a basis for our examination of the behaviour of nuclei, which forms the second theme of this book.

Rutherford's Laboratory in Cambridge, 1933.

The Antiproton Accumulator at CERN, the European collaborative laboratory at Geneva.

Chapter 1

ATOMS AND NUCLEI

LEARNING OBJECTIVES

After studying this chapter you should be able to:

1. name the constituent particles of nuclei, and state the typical dimensions of nuclei;

2. describe the evidence from alpha particle scattering and electron diffraction experiments;

3. show how the constancy of the density of nuclear material derives from such evidence.

1.1 SOME FAMILIAR PARTICLES

Introduction

You are no doubt aware that matter is made of atoms. The idea that matter is not infinitely divisible, but is made up of particles called atoms, has been around for a long time and is now generally accepted as a useful description. You will also be aware that it is possible to identify particles which make up the atom – protons, neutrons and electrons. The idea that scientists can split the atom is now well known and has impressed itself on the public consciousness, particularly through developments in nuclear power and nuclear weapons. But did you know that the word 'atomic' means 'unsplittable'? Originally atoms were thought of as just that: unsplittable, the smallest possible particles.

To understand how scientists have developed our present picture of the atom, with electrons orbiting a central nucleus, we must think back to the state of atomic theory at the turn of the twentieth century. Scientists working with cathode ray tubes had identified positively and negatively charged particles. In a famous experiment, J J Thomson had succeeded in measuring the charge-to-mass ratio of cathode rays. He identified these rays as beams of electrons. You have probably seen or carried out a similar experiment to determine this ratio, e/m.

However, the positive rays were found to have much smaller charge-to-mass ratios. Since they were expected to have a charge equal and opposite to the electron's charge, it followed that they must be many times more massive than the electron.

Table 1.1 lists the masses and charges of the three particles which we will consider in our picture of the atom. (The neutron is included although this was not discovered until the 1930s.) The table shows the measured values of mass and charge, both in SI units and relative to those for the electron.

Fig 1.1 J J Thomson demonstrating the measurement of e/m.

Table 1.1 Values of charge and mass for three subatomic particles.

	Charge/C	Mass/kg	Charge/e	Mass/m_e
electron	$- 1.602 \times 10^{-19}$	9.110×10^{-31}	$- 1$	1
protron	$+ 1.602 \times 10^{-19}$	1.673×10^{-27}	$+ 1$	1836.1
neutron	0	1.675×10^{-27}	0	1838.6

The plum pudding model

From the knowledge gained from these experiments, Thomson devised a model of the atom known as the plum pudding model. We now know that this model was wrong, and we will look at Rutherford's experiment which disproved it in Section 2.2. However, in order to understand later work, particularly Rutherford's experiment, we should briefly examine Thomson's model.

Thomson argued that, since matter in bulk is uncharged, atoms themselves must be uncharged. He pictured the atom as being made up of equal amounts of positively and negatively charged matter. Electrons have very little mass, and so most of the mass must be concentrated in the positive matter.

But how might the charge and mass be distributed within the atom? Thomson argued that, because the positive and negative charges attracted and balanced one another in a neutral atom, they must both be distributed throughout the atom. He pictured the negative charge, in the form of electrons, distributed within the massive, positive charge which made up the bulk of the atom.

This is the plum pudding model, as shown in Fig 1.2(a). Electrons form the plums within a pudding of positive charge. This model is both simple and plausible, and it provided a satisfactory explanation of the data available to Thomson at the time. He had little information on the size of subatomic particles, and the existence of the neutron was unknown.

It would have been difficult for Thomson and his colleagues to imagine that all the positive charge within the atom might be concentrated in a very small space, as in the nuclear model of the atom shown in Fig 1.2(b). After all, like charges repel one another, and so it seemed obvious that the positive charge should be spread throughout the atom, with the negative electrons distributed throughout to counteract this electrostatic or Coulomb repulsion.

We will return to the plum pudding model when we examine Rutherford's experiment shortly. In the meantime, to help our discussion, we should look at the ways in which the compositions of different nuclei are represented, as well as the energy units used in nuclear physics.

positively charged sphere

(a)

(b)

Fig 1.2 Two models of the atom.
(a) In the plum pudding model, the electrons are distributed within the positive charge which makes the bulk of the atom.
(b) In the nuclear model, electrons orbit a central, positively charged nucleus.

Nuclear notation

It is useful to have a simple notation to describe the composition of a particular nucleus, or **nuclear species**. Nuclei differ in the numbers of protons and neutrons they contain. Protons and neutrons are collectively known as nucleons. The conventional symbols for these quantities are:

$$Z = \text{number of protons}$$

$$N = \text{number of neutrons}$$

$$A = \text{number of nucleons}$$

Since the protons and neutrons together comprise the nucleons, it follows that:

$$A = Z + N$$

A particular nuclear species of an element X is written as:

$$_Z^A\text{X} \quad \text{or} \quad _Z^A\text{X}_N$$

The first of these notations will generally be used in this book, since, given Z and A, you can simply calculate N. Here are some examples of nuclei represented in this way:

$$_1^1\text{H} \quad _2^4\text{He} \quad _6^{14}\text{C} \quad _8^{16}\text{O} \quad _{92}^{235}\text{U} \quad _{17}^{35}\text{Cl} \quad _{17}^{37}\text{Cl}$$

Note that the last two are the commonest forms of the element chlorine. They have the same number of protons (17). These two forms are referred to as **isotopes** of chlorine.

Incidentally, don't be surprised if you come across other notations, such as U-235 or carbon-14. The number is the value of A. Knowing the value of Z for the element concerned, you should be able to translate these into the more standard form above.

Nuclear energies

The SI unit of energy, the joule, is convenient for energies on a macroscopic scale. If you climb to the top of a ladder, the increase in your gravitational potential energy is of the order of 1000 joules. If you eat a cheese sandwich, it gives you about one megajoule of energy. However, these energies are far greater than any encountered on the scale of an individual nucleon.

For convenience we need to use a smaller unit, the **electronvolt**, for which the symbol is eV. If you are familiar with this unit, and how to convert between joules and electronvolts, test yourself on the questions at the end of this section.

One electronvolt is defined as the energy transformed when an electron moves through a potential difference (pd) of one volt. Since energy = charge × potential difference, and the charge on an electron, e, is about 1.6×10^{-19} C, it follows that:

$$1 \text{ eV} = 1.6 \times 10^{-19} \text{ J}$$

To convert from eV to J, multiply by e.

To convert from J to eV, divide by e.

For example:

1. An electron moves through a pd of 2000 V. Its energy is, therefore, 2000 eV, or 2 keV. In joules, this is $2000 \times e$, or 3.2×10^{-19} J.

2. A car of mass 750 kg travelling at 20 ms^{-1} has 150 000 J of kinetic energy. In eV, this is $150\,000/e$, or 9.375×10^{23} eV.

You should be able to see from these examples that, for convenience, we use the appropriate units to avoid having large powers of 10 to handle. But bear in mind that the equations we will be using later assume that quantities are in SI units, so you must be able to convert between eV and joules.

QUESTIONS	
	1.1 An alpha particle emitted in the radioactive decay of uranium-238 has energy 4.19 MeV. How many joules is this?
	1.2 An electron is accelerated through a cathode–anode pd of 1 kV. What is its kinetic energy in eV, and in J? If its mass is 9.11×10^{-31} kg, what is its final velocity?
	1.3 Find the energy, in eV, of a photon of red light of energy 2.8×10^{-19} J.

1.2 RUTHERFORD'S ALPHA SCATTERING EXPERIMENT

Introduction

In physics, if we want to know about the structure of something, we often shoot things at it and see what happens. We may use beams of particles or energy. To investigate the structure of solids, we use X-rays; to find out about the interior of the Earth, we use sound waves.

In Rutherford's day, there were no particle accelerators providing a ready supply of particle beams, and so he had to look around for a natural probe which might yield information about atomic structure. Alpha particles are a natural product of many radioactive substances, and their properties had been under investigation for over a decade. To investigate atomic structure, Rutherford decided to try alpha particles. (You should recall that an alpha particle consists of two protons and two neutrons; it therefore has charge $+2e$. It is the same as the nucleus of a helium atom, ^4_2He.)

An atom is, of course, too small to be investigated individually, so Rutherford decided to direct a beam of alpha particles at a metal foil consisting of many atoms. The nature of individual atoms had then to be inferred from the observed scattering of the beam.

Experimental details

Working with Geiger and Marsden, Rutherford devised an experiment to investigate the scattering of a beam of alpha particles by a thin foil of gold. The experimental arrangement is shown schematically in Fig 1.4.

Fig 1.3 Ernest Rutherford (right) had a very loud voice, which sometimes interfered with the sensitive counting instruments in his laboratory.

TOP VIEW

Fig 1.4 The construction of Rutherford's alpha particle scattering experiment.

The idea of the experiment was to determine the extent to which alpha particles were deflected as they passed through the gold foil. A radioactive source (Rutherford used radon gas) emits radiation equally in all directions. Therefore it was first necessary to produce a narrow (collimated) beam. Without a narrow beam it would have been impossible to determine accurately the angle through which the particles were deflected.

Gold was used, since it is easy to roll out a foil which is only a few atoms thick. The deflected alpha particles were detected by a phosphor screen. Particles striking the screen made the phosphor glow with flashes of light, and these were counted to determine the numbers of particles deflected through different angles. The experiment had to be performed in a vacuum chamber, since alpha particles are absorbed by a few centimetres of air at atmospheric pressure.

Fig 1.5   Geiger and Rutherford in their laboratory at Manchester University.

Fig 1.6   Rutherford's apparatus (1914) for measuring the charge to mass ratio of alpha particles.

gold atom

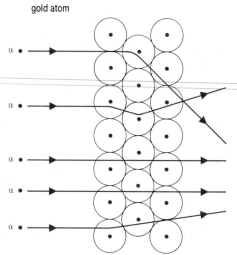

Fig 1.7   Typical paths of alpha particles striking a gold foil. Only those whose paths bring them close to a nucleus are significantly deflected. (Note that the nuclei, if drawn correctly to scale, would not be visible on this diagram.)

Geiger and Marsden's results

Geiger and Marsden found that most alpha particles were deflected through only small angles. This might have been expected, since alpha particles are fast-moving (with speeds of the order of 10^7 m s^{-1}) and their masses are comparable to atomic masses. Therefore they might not be expected to be significantly deflected as they passed through the gold foil. Indeed, most of the alpha particles which were passed through the foil scarcely deflected at all.

However, the most striking feature of the results obtained by Geiger and Marsden is that, when they moved the detector round close to the alpha source, they observed a small but significant number of flashes of light. This meant that some alpha particles were being scattered back towards the source. This phenomenon, where a particle is scattered through an angle greater than 90°, is called **back-scattering**; about one in 8000 alpha particles were found to be back-scattered.

Rutherford's analysis

Rutherford was very struck by the phenomenon of back-scattering. He said that it was as if one had fired a bullet at a sheet of tissue paper, and the bullet had bounced back. Certainly, the plum pudding model of the atom could not explain this observation.

Imagine building a wall of plum puddings, a few puddings thick. Now imagine firing a fusillade of lead shot at it. You would probably feel safe from rebounding shot! However, suppose your cooking had gone wrong and each pudding dough had a very dense, solid core at its centre. Now you might be in danger of being hit by reflected lead shot.

This was Rutherford's explanation of the alpha particle experiment. He said that the positive charge and most of the mass of the atom were concentrated into a small region of space within the atom – the nucleus. The region outside the nucleus would then be occupied by the electrons.

Fig 1.7 shows the paths of typical alpha particles striking the foil. Since the foil is only a few atoms thick, and the nucleus occupies only a tiny fraction of the atomic volume, most alpha particles do not interact significantly with the nucleus. The electrons, because their mass is small,

ATOMS AND NUCLEI

have little effect; the momentum of the alpha particles carries them through the foil.

Those alpha particles which happen to pass close to a nucleus experience electrostatic repulsion. Since the mass of a gold nucleus is many times that of an alpha particle, the alpha particle may be deflected significantly. The gold nucleus recoils slightly. An alpha particle whose trajectory brings it into head-on collision with a gold nucleus is scattered back towards the source.

While back-scattering is the most striking observation in this experiment, and in itself provides strong evidence for the nuclear model of the atom, Rutherford's analysis of the results was much more detailed than this. Geiger and Marsden measured the proportions of alpha particles scattered through different values of the scattering angle θ (defined in Fig 1.9). From this, and assuming that the nucleus was spherical, it was possible to calculate the approximate diameter of the nucleus. The results obtained were so detailed that Rutherford was able to conclude that Coulomb's Law (the inverse square law of electrostatic repulsion) was valid down to distances of the order of 10^{-14} m, distances much smaller than any atomic dimensions.

INVESTIGATION

Fig. 1.8   A model of Rutherford scattering.

Fig 1.9   Diagram to show the aiming error p and the scattering angle θ between the incoming and scattered paths.

Fig 1.8 shows an experiment in which the scattering of alpha particles by a nucleus is modelled by the scattering of a ball-bearing from a plastic or metal dome. In this way it is possible to see the different tracks traced out by alpha particles approaching a nucleus with different degrees of closeness.

In Fig 1.8 a ball-bearing runs down the small ramp and is deflected as it passes the plastic dome. The closer it comes to a head-on collision, the greater the angle through which it is deflected. The tracks may be shown up by sprinkling lycopodium powder on the dome. You can make your own dome from plasticine, if you do not have a plastic one.

Alternatively, a magnetic air puck, pushed gently towards a stationary magnetic object, is scattered through an angle which depends on its original line of approach.

Try out one or other of these experiments. Whichever you choose as your model of alpha scattering, you should be able to devise procedures to enable you to answer questions 1.4 and 1.5 which follow.

There are two important measurable quantities in this experiment. They are the scattering angle, θ, and the aiming error, p. These are shown in Fig 1.9, which describes the experiment schematically, as viewed from above. You must devise techniques for varying p, and for measuring both p and θ. Record how you have achieved this.

Fig 1.10   Tracks of alpha particles passing close to an atomic nucleus.

1.4 How does the angle θ change as you increase the aiming error p? Devise a method for varying the speed v at which the incoming ball-bearing or puck approaches the scatterer. For a fixed value of p, investigate how θ depends on v.

1.5 Does θ increase or decrease with increasing v?

Your account of this investigation should include details of how you have varied p and v, and how you have measured θ. Include a sketch to show typical tracks of scattered particles. Fig 1.10 shows the corresponding picture for alpha particles.

The size of the nucleus

We do not need to go into Rutherford's analysis in depth to get an idea of the nuclear size. We can simply use the fact that back-scattering is indeed observed to get a crude estimate of the nuclear diameter.

We have to think about the energy of an alpha particle as it is scattered by a gold nucleus. Imagine a head-on collision, as an alpha particle is directed straight at the nucleus. This is shown in Fig 1.11.

Initially the alpha particle has kinetic energy. It gradually experiences electrostatic repulsion due to the positively charged nucleus. It slows down as its kinetic energy changes to electrostatic potential energy. At its point of closest approach to the nucleus, it comes instantaneously to rest; at this moment its initial kinetic energy has been converted entirely to potential energy. Then the process reverses, as the alpha particle is scattered back towards the source.

(In the model of Fig 1.8 the kinetic energy of the ball-bearing is converted to gravitational potential energy as it runs up the slope of the dome.)

We can equate the kinetic energy (E_k) of the alpha particle with the potential energy (E_p) it has at its nearest point to the nucleus. From this we can deduce the distance of closest approach, and this will give us some idea of the scale of the nucleus.

The potential energy of an alpha particle, charge Z_α, in the Coulomb field of a gold nucleus, charge Z_{Au}, is given by:

$$E_p = \frac{1}{4\pi\epsilon_0} \times \frac{Z_\alpha Z_{Au}}{r}$$

where r is the centre-to-centre separation of the two particles, and ϵ_0 is the permittivity of free space (see Appendix C). At the point of closest approach the initial kinetic energy of the alpha particle has been converted to potential energy, and so we can write:

$$E_k = \frac{1}{4\pi\epsilon_0} \times \frac{Z_\alpha Z_{Au}}{d}$$

where d is the distance of closest approach.

Fig 1.11 Energy changes as an alpha particle scatters from a gold nucleus.

Now you can use what you know about these particles to find d.

1.6 The kinetic energies of alpha particles are typically about 5 MeV. How much is this in joules?

1.7 The charge on an alpha particle is +2e, and that on a gold nucleus is +79e. Use the equation above to find the distance of closest approach, d.

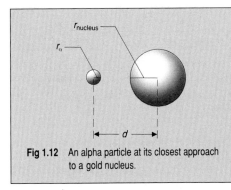

Fig 1.12 An alpha particle at its closest approach to a gold nucleus.

Now that you have obtained a value for *d*, what does this tell us about the size of the nucleus?

Fig 1.12 shows the two particles at the instant of closest approach. Note that we cannot say they are touching; indeed, this may not have any meaning for subatomic particles. We should not fall into the trap of thinking of these particles as being just like the hard macroscopic particles which we are used to experimenting with in the laboratory.

1.8 What can we conclude about the radius of the gold nucleus? How does it compare with the value of *d*?

Conclusions

It is now over eighty years since Rutherford provided this clear evidence in support of the nuclear model of the atom. His results gave the first good indication that the nucleus occupies only a tiny fraction of the volume of the atom. Most of the mass of the atom, and all of its positive charge, is concentrated in this small volume. Whilst alpha scattering experiments have given valuable information, other techniques have become available for probing the nucleus. In the rest of this chapter we will look at some of these techniques and try to understand what they can tell us about the nature of the nucleus.

QUESTIONS

1.9 Think about the alpha scattering experiment. Why must the alpha particles be collimated to give a very fine beam?

1.10 What would you expect to be the effect of using a source of higher energy (faster) alpha particles on the fraction of particles back-scattered?

1.3 ELECTRON DIFFRACTION

A question of wavelength

You should be familiar with the basic ideas of the diffraction of waves. When waves pass through a gap in a barrier, or when they travel past the edge of an obstacle, they are diffracted; that is, they tend to spread out into the space beyond the barrier. The biggest diffraction effects are observed when the wavelength of the waves is about the same as the width of the gap.

This phenomenon is made use of in many experimental techniques. You will have used diffraction gratings to observe diffraction effects using visible light. Given the slit spacing of the grating, we can find the wavelength of the light.

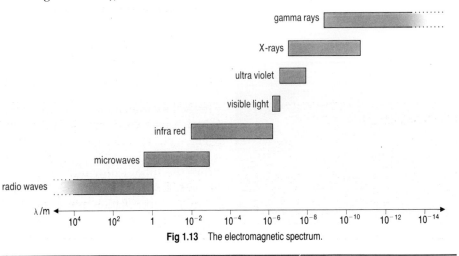

Fig 1.13 The electromagnetic spectrum.

Another example of the use of waves is X-ray crystallography, in which X-rays are diffracted by the parallel planes of atoms which make up a crystal. The atomic planes act like a diffraction grating. This time the calculation is done in reverse; given the wavelength of the X-rays, we can find the separations of the atomic planes.

In both cases radiation is used which has a wavelength appropriate to the physical dimensions of the diffracting system. In the case of visible light the wavelength is about 500 nm, and the slit separation of diffraction gratings such as you may have used in the laboratory is typically 3 μm, a few times longer than the wavelength. In X-ray crystallography the separation of atomic planes is of the order of 0.1 nm, and visible light is not suitable. X-rays are used instead, as they are electromagnetic waves of appropriate wavelength.

(If you are not certain about these basic ideas of diffraction, you should check them out in a standard physics textbook [see Appendix A].)

What can we use as a suitable tool for investigating nuclei? As discussed in Section 1.2 nuclear dimensions are smaller than 5×10^{-14} m. We might consider the appropriate electromagnetic radiation. Look at the electromagnetic spectrum shown in Fig 1.13. X-rays clearly have too long a wavelength to be of use.

QUESTION

1.11 From Fig 1.13, which electromagnetic radiation might we consider using to investigate nuclei?

Particles and waves

Gamma rays have a shorter wavelength than X-rays and also originate within the nucleus, so we might have anticipated that they would be an appropriate tool for investigating the nucleus. However, it turns out in practice that it is not very easy to produce a beam of gamma rays and therefore they are not a convenient way to find out about the nucleus.

Given that electromagnetic radiation is unsuitable, we must turn our attention to beams of particles. Several different particle beams have been used in nuclear research; we will give our closest attention to the diffraction of electron beams.

We are used to thinking of electrons as particles. They are matter; they have mass, and they have charge. We can explain much of their behaviour, for example in a cathode ray tube or as part of an electric current in a wire, by describing electrons as particles. However, when a beam of fast-moving electrons interacts with solid matter, we have to adopt a different view of the electron in order to interpret the observed results. We picture the electrons as having wave properties; they are diffracted as they pass through matter.

This phenomenon is called **wave–particle duality**. Sometimes we must interpret the behaviour of electrons as wave-like, and at other times as particle-like. The same is true for visible light. It can be shown to diffract – a wave phenomenon. At other times, for example in the photoelectric effect, we interpret its behaviour in terms of particles called photons.

In the laboratory, electron diffraction can be shown by passing a beam of cathode rays (electrons) through a thin piece of graphite. This is illustrated in Fig 1.14. The electrons are diffracted by the atomic planes within the graphite, and diffraction rings are observed on the fluorescent screen at the end of the tube. Knowing the cathode–anode pd (typically a few kV), we can calculate the electron velocity, and hence the wavelength. Measuring the angle of diffraction allows us to calculate the spacing of atomic planes in graphite.

This equipment can be used to determine the spacing of atomic planes,

Fig 1.14 A demonstration of electron diffraction.

ATOMS AND NUCLEI

of the order of 10^{-10} m, for example in graphite. But we want to investigate the nucleus, which is about 10^5 times smaller. We must change the electron beam to make it a suitable probe for the nucleus.

What must we do? A clue comes from the **de Broglie equation**. This relates a particle property, momentum p, to a wave property, the wavelength λ:

$$p = \frac{h}{\lambda}$$

(Here, h is Planck's constant.) This is the fundamental equation of wave–particle duality. It allows us to translate between the wave and particle behaviours of matter. We will use it now to think about the way in which we can produce a beam of electrons with a wavelength sufficiently short to probe the atomic nucleus.

ASSIGNMENT

We want to reduce the wavelength of the electrons to about 10^{-15} m. Consider the de Broglie equation, and use it to answer the questions which follow.

1.12 What must we do to the momentum of the electrons?

1.13 What effect will this have on their energy?

1.14 What must we alter in the experimental arrangement of Fig 1.14 to have this effect?

To reach suitably small wavelengths, the energetic electrons we require have a velocity approaching the speed of light, c. We say that they are relativistic, since we must use the equations of relativity. (Newton's laws of motion do not apply for such fast-moving particles.) The energy E of a relativistic electron is related to its wavelength λ by the equation:

$$E = \frac{hc}{\lambda}$$

(Note that this is another equation translating between wave and particle properties.)

1.15 Use this equation to find the energy required for electrons to have a wavelength of 10^{-15} m. You should calculate the value of E both in joules and in MeV.

1.16 What cathode–anode pd is needed to accelerate electrons to such an energy?

Diffraction patterns

Beams of high energy electrons are produced in practice using large linear accelerators, based on the same principle as the van de Graaff generator. High voltages, of the order of hundreds of MeV, may be reached; given that it is not feasible for you to carry out such an experiment, we will look at some simple analogue experiments, before considering the results found using linear accelerators.

INVESTIGATION

Two simple diffraction experiments are shown in Fig 1.15. From these experiments you will obtain diffraction patterns for light being diffracted by a ball-bearing, and by a random array of lycopodium dust particles.

Fig 1.15 Two diffraction experiments.

Fig 1.16 Two diffraction patterns.

In Fig 1.15(a) a laser beam is diffracted by a small, spherical ball-bearing mounted on the head of a pin. **Safety note**: do not look into the laser beam.

In Fig 1.15(b) lycopodium dust is sprinkled thinly on a microscope slide and a small lamp viewed through the slide.

Try these experiments now. Sketch the diffraction patterns you observe. How does the pattern for a single ball-bearing differ from that for a random array of lycopodium dust particles? How do these patterns compare with that for electron diffraction shown in Fig 1.14?

Fig 1.16 shows the two diffraction patterns corresponding to the two experiments in Fig 1.15.

Electron diffraction

Of course, we cannot balance a single nucleus on the head of a pin. We have to direct our electron beam at a sample containing many millions of nuclei. The diffraction patterns obtained in electron diffraction are much more similar to the blurred rings obtained in the lycopodium experiment shown in Fig 1.16(b) than those of the ball-bearing experiment.

Let's extend our analysis of these diffraction patterns a little further. We observe rings of varying intensity. The pattern is bright at the centre, then dimmer, then bright again as we cross the first ring, then dim again, then bright again at the second ring. This is represented in Fig 1.17(a) as a graph of intensity against the angle of diffraction, θ. (θ is the angle through which the beam has been diffracted.) Compare this with Fig 1.17(b) – an intensity-angle graph obtained from an experiment in which a beam of high-energy electrons have been diffracted by carbon. Although the two graphs are not identical, you should be able to trace in Fig 1.17(b) the ups and downs of the diffraction pattern represented in Fig 1.17(a).

Diffraction theory allows us to deduce from these patterns the size of the diffracting objects. Identify the angle in Fig 1.17(b) at which the first

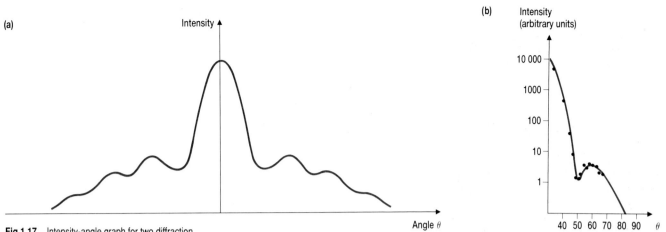

Fig 1.17 Intensity-angle graph for two diffraction patterns.
(a) Visible light diffracted by lycopodium dust particles;
(b) High-energy (420 MeV) electrons diffracted by carbon nuclei.

diffraction minimum occurs – this is the dip at about 51°· The radius r of the diffracting nuclei is related to this value of θ by the equation:

$$r = \frac{0.61\ \lambda}{\sin\ \theta}$$

(This equation assumes that the diffracting object is a hard sphere.) Using this equation you can estimate the value of r for the carbon nucleus.

QUESTIONS

1.17 Calculate λ for electrons of energy 420 MeV.

1.18 Use the equation above to find r, and hence the diameter of the carbon nucleus.

Models of the nucleus

We have seen how electron diffraction can tell us about the size of the nucleus. However, much more information can be derived from diffraction patterns by careful analysis of graphs such as Fig 1.17(b). Any such interpretation depends on which model of the nucleus is used as a basis for analysis. For example, the previous discussion assumed the nucleus to be a 'hard sphere' – that is, a positively charged sphere of uniform charge density – and having a clearly defined surface. A more sophisticated model might assume that the spherical nucleus had a 'fuzzy' surface, with the density of charge gradually decreasing outwards from the centre of the sphere. We need not concern ourselves with such models, but you should bear in mind that any data you see giving values for nuclear dimensions have been deduced on the basis of particular theoretical models of the nucleus. The nucleus is neither a hard sphere nor a fuzzy one; these are simply models which allow us to achieve a reasonable understanding of experimental data.

Other nuclear probes

Electron diffraction is not the only tool available to nuclear physicists seeking details of nuclear dimensions. We discussed alpha particle scattering in Section 1.2; in practice, the alpha particles produced by natural radioactive sources do not have enough energy to get very close to the nucleus. They must be accelerated if they are to overcome the strong Coulomb (electrostatic) repulsion of the nucleus.

Other particle beams used to probe the nucleus include protons and neutrons. These are both constituent particles of nuclei; they experience nuclear forces as well as (in the case of protons) Coulomb repulsion. As a consequence, they give information about the nucleus which is different from that derived from electron diffraction. Electrons only experience the Coulomb force, and so they give information about the distribution of charge within the nucleus. This is an example of the way in which different experimental techniques provide different information about the target of our investigations.

In the next section we will look at some of the information on nuclear dimensions provided by these techniques, and then go on to deduce what we can about the nature of the nucleus.

QUESTIONS

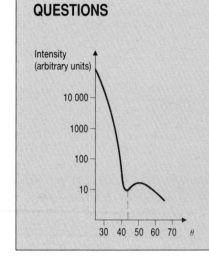

Fig 1.18 shows an intensity–angle plot for the diffraction of 420 MeV electrons by oxygen.

1.19 Use the graph to deduce a value for the diameter of the oxygen nucleus.

1.20 Which do you think would penetrate closer to the nucleus, a beam of high energy protons, or a beam of equally energetic neutrons? (Hint: think about the forces acting on the particles.)

Fig 1.18 The variation of intensity with angle of diffraction for high energy electrons striking an oxygen target.

1.4 THE NUCLEAR RADIUS

Results and data analysis

In Table 1.2 you will see data derived from electron diffraction experiments on several different elements. The table lists values of the nuclear volume and radius, derived assuming the nucleus to be a uniform sphere of positive charge. We will analyse this data in the following assignment.

Table 1.2   Values of nucleon number A, nuclear volume V and radius r for several nuclear species.

Element	Symbol	$V/10^{-45}\ \mathrm{m}^{-3}$	$r/10^{-15}\mathrm{m}$	A
carbon	C	132	3.16	12
silicon	Si	254	3.93	28
iron	Fe	478	4.85	56
tin	Sn	900	5.99	120
lead	Pb	1538	7.16	208

ASSIGNMENT

1.21 Look at the data in Table 1.2. How does the nuclear volume V vary as the nucleon number A increases?

In Fig 1.19 the variation of nuclear radius r is plotted against A. From this graph it is clear that r increases uniformly with A. The more nucleons there are in the nucleus, the greater the observed nuclear radius. This might not be surprising but it is an important observation. It tells us something about the nature of the forces which hold the nucleus together. We will discuss this further in the next section.

There is a mathematical relationship between A and r which we can guess at as follows: we might expect the volume V of the nucleus to be proportional to the number of nucleons A. But V is also proportional to

r/10⁻¹⁵ m

Fig 1.19 The nuclear radius increases uniformly with the number of nucleons, *A*: data from Table 1.2.

the cube of the radius *r*. So we might expect to find $A \propto r^3$. Inverting this relationship, we can write:

$$r = r_0 A^{1/3}$$

In words, this says that the radius *r* is proportional to the cube root of the number of nucleons. By plotting a suitable graph, you can find the value of the constant of proportionality, r_0.

1.22 Compare the equation for *r* with the equation for a straight line, $y = \mathrm{m}x + \mathrm{c}$. What should you choose for the quantities on the *x*- and *y*-axes of a graph to obtain a straight line? What quantity will the gradient tell you the value of? What would you expect the intercept on the *y*-axis to be? Draw up a suitable table of values, derived from those in Table 1.2. Plot a graph using these values.

If you have obtained a straight line graph, you have succeeded in confirming the relationship $r \propto A^{1/3}$. You may find that the line does not pass exactly through the origin; the data do not fit the relationship perfectly for small values of *A*.

1.23 From your graph, deduce a value for r_0. We might take this as an approximate value for the radius of the proton since, when *A* = 1, the equation becomes $r = r_0$.

A simple calculation

We can calculate the density of nuclear matter using the results of the previous assignment.

Picture the nucleus as a sphere of radius r, mass M and density ρ. It is made up of A nucleons, each of mass m. The mass of the nucleus can be written in two ways. Firstly, M is the volume of the spherical nucleus times its density:

$$M = \tfrac{4}{3}\pi r^3 \rho$$

Since $r = r_0 A^{1/3}$, we can write $r^3 = r_0^3 A$. Hence:

$$M = \tfrac{4}{3}\pi r_0^3 A\rho$$

Secondly, M is approximately equal to the total mass of the constituent nucleons:

$$M = Am$$

Combining these two gives:

$$Am = \tfrac{4}{3}\pi r_0^3 A\rho$$

Notice that there is a factor of A on both sides of this equation, so these cancel. Rearranging to find an expression for ρ gives:

$$\rho = \frac{3m}{4r_0^3}$$

From this expression, you can calculate a value for the density of nuclear matter.

ASSIGNMENT

Refer to the value for r_0 which you deduced in question 1.23. The value of m is approximately the mass of a proton, 1.67×10^{-27} kg.

The atomic nucleus is enormously dense; for comparison, the density of water is 10^3 kg m^{-3}. The difference arises because, of course, the nucleus occupies only a tiny fraction of the volume of the atom. The rest of the atom is occupied by the electrons, which contribute very little to the atomic mass.

1.24 What value do you obtain for the density of nuclear matter? How much denser is this than water?

The constancy of r

Since the expression for r does not involve A, it follows that the density of nuclear material is constant; all nuclei have more or less the same density.

(This is not the case for all atoms. Different atoms have different densities, and hence different elements have different densities. For example, an atom of gold is almost exactly the same size as an atom of silver, and yet the mass of a gold atom is almost twice that of a silver atom. The density of an atom of gold is therefore almost twice that of a silver atom. Their nuclei, however, have the same density.)

$A \sim 20$ $A \sim 50$ $A \sim 100$ $A \sim 200$

Fig 1.20 As nuclei contain progressively more nucleons, they become progressively larger.

The density of nuclear material does not vary much from one nuclear species to another. This is what we observed in the previous section; as the value of A increases, the nuclear radius increases steadily.

This gives us an idea of how we can picture different nuclei. Imagine building up progressively bigger nuclei (see Fig 1.20). We start with the simplest nucleus, a single proton. Add another nucleon, which has about the same mass and volume as the proton. The mass and volume are doubled, so the density remains constant. Add more nucleons, and the density is virtually unaltered.

This is often likened to the way in which a drop of water may be built up from individual water molecules. As more molecules are added to the drop, the volume and mass of the drop increase together, so that the density remains constant. The molecules retain their individual identities. In the next chapter, we will use this **liquid drop model** of the nucleus to give us some more hints about the nature of forces within the nucleus.

1.6 NUCLEAR SCALES

The nucleus within the atom

We finish this chapter by summarising the picture we have built up so far of the nature of the atom.

Atoms are typically of the order of 10^{-10} m across. At the centre is the tiny nucleus, about 10^{-15} m across. The nucleus thus occupies about one part in 10^{13} or 10^{14} of the atomic volume, but it accounts for all but one part in 5000 of the atomic mass. It contains all the positive charge of the atom. This charge is largely shielded from the nuclei of neighbouring atoms by the surrounding electrons. There is little influence exerted by one nucleus on neighbouring nuclei. This is why radioactive decay is a spontaneous process; one nucleus cannot sense when another has decayed.

Because the nucleus is so small and the electrons have so little mass, when an alpha particle passes through a metal foil it is unlikely to be deflected unless it should chance to pass close to a nucleus.

Units

Because the dimensions of the nucleus are so much smaller than those of the macroscopic world in which we live, it is usual to use units other than SI units for many quantities. We have already discussed the electronvolt as a unit of energy; nuclear energies are typically measured in MeV. Particle accelerators produce particles with energies in the GeV range (Remember: $1\,\text{GeV} = 10^9\,\text{eV}$).

Linear dimensions on the nuclear scale are typically multiples of 10^{-15} m. The appropriate SI prefix is f, which stands for femto ($1\,\text{fm} = 10^{-15}$ m). You may see this referred to as the **fermi**, after Enrico Fermi, who made many important contributions to the development of nuclear physics. (The fermi is not an SI unit.)

We can get an idea of the time-scale of nuclear processes as follows. The nuclear diameter is about $1\,\text{fm}$. The fastest speed with which information could be transmitted across the nucleus is the speed of light in free space, $c = 3 \times 10^8\,\text{m s}^{-1}$. The time taken for light to cross the nucleus is thus about $10^{-15}/3 \times 10^8$ s, or about 3×10^{-24} s. This gives us a general idea of the time-scale on which we might expect nuclear processes to happen.

Masses are frequently expressed in unified atomic mass units, the symbol for which is u. On this scale, the mass of a proton or neutron is approximately $1\,\text{u}$. This unit is defined by taking the mass of one atom of ^{12}C as exactly $12\,\text{u}$.

If you are not sure about the definition of the unified atomic mass unit, and the way to convert between masses in u and masses in kg, try the questions which follow.

1.25 12 g of ^{12}C contains Avogadro's number (6.022×10^{23}) of nuclei. Show that 1 u is approximately 1.661×10^{-27} kg.

1.26 The mass of an atom of ^{235}U is 235.0439 u. Convert this to kg, correct to six significant figures.

1.27 1 kg of ^{14}C contains 4.3004×10^{25} atoms. Calculate the mass of one atom, in kg and in u.

To sum up....

The nuclear model of the atom is now accepted as a good description of the distribution of matter within the atom. Since the nucleus is so much smaller than anything of which we can have direct experience in daily life, exotic experimental techniques are needed to elucidate many of its features, and exotic theories are needed to explain the results of such experiments.

Don't be surprised if some of the things that go on within the nucleus seem strange at first. It can take an imaginative leap to get to grips with them. Perhaps the most remarkable thing is that, within eight decades of Rutherford's demonstration of the existence of the nucleus, we have come to know so much about its inner workings.

SUMMARY

The existence of a nucleus within the atom was deduced by Rutherford from alpha particle scattering experiments. Most of the mass and all of the positive charge of an atom are concentrated in the nucleus, which occupies a very small fraction of the atomic volume, 1 part in 10^{14}. The size of the nucleus is determined by electron diffraction (and other techniques). The volume of the nucleus increases in proportion to the number of nucleons, A; all nuclei have approximately the same density.

Chapter 2

FORCES IN THE NUCLEUS

Many nuclei are very stable; many of the nuclei which formed in the early moments of the life of the universe are still in existence after 15 000 000 000 years. What powerful forces are there which can hold these nuclei together? (See Fig 2.1.)

LEARNING OBJECTIVES

After studying this chapter you should be able to:

1. state the relative importance of the strong nuclear, electrostatic and gravitational forces in holding nucleons together in a nucleus;

2. describe the characteristics of the strong nuclear force between nucleons;

3. relate force-separation and energy-separation graphs for the strong nuclear force.

(a)
the Sun

(b)
an atom

(c)
a nucleus

Fig 2.1 (a) Gravity holds a star together.
(b) Electrostatic attraction holds an atom together.
(c) So what holds a nucleus together?

2.1 GRAVITATIONAL AND COULOMB FORCES

Introduction

In our discussion of the composition of the nucleus you may well have noticed a problem; the nucleus contains positively charged particles, protons, which must repel one another according to Coulomb's Law. And yet many nuclei are very stable; they do not burst apart under the influence of these repulsive forces between protons.

An atom, of course, is electrically neutral. It contains equal numbers of protons and electrons, and their charges balance. The nucleus, however, contains no negatively charged particles. The protons are closely packed together in a very small volume. Because Coulomb's Law is an inverse square law, we might expect to find that the protons, being so close together, would exert very strong repulsive forces on each other.

Consider the forces acting between two neighbouring protons in a nucleus. The electrostatic force F_e which a charge Q_1 exerts on another charge Q_2 (at a distance r) is given by:

$$F_e = \frac{1}{4\pi\epsilon_0} \times \frac{Q_1 Q_2}{r^2} \qquad (1)$$

An attractive force which you are probably familiar with is gravitation. Every particle which has mass attracts every other particle according to Newton's Law of Gravitation, another inverse square law. The gravitational force F_g exerted by a mass m_1 on another mass m_2 is given by:

$$F_g = -G \frac{m_1 m_2}{r^2} \tag{2}$$

Here we have one attractive force and one repulsive force acting between protons in a nucleus. Is the gravitational attraction between them sufficient to overcome the electrostatic repulsion?

ASSIGNMENT

Consider two protons in a nucleus, separated by a distance of 1fm (the sort of distance between two nucleons in a nucleus). You can now calculate the two forces between these particles.

2.1 Use equation (1) to calculate F_e. Do you consider that your answer represents a large force on a proton? (You can get an idea of how big a force this is for a proton by calculating the acceleration it would experience as a result of this force, using $F = m\,a$.)

2.2 What value do you find for the acceleration a?

F_e is indeed a large force on the scale of a proton – it is roughly the weight of a small child, exerted on a very small particle. With each proton in the nucleus exerting such a large force on each of the other protons, we might well expect the nucleus to be very unstable.

Now use equation (2) to determine whether the attractive force of gravity between two neighbouring protons in the nucleus is enough to balance their electrostatic repulsion.

2.3 What value do you find for F_g?

2.4 What is the ratio of electrostatic repulsion to gravitational attraction?

Another force?

The assignment shows that gravity is negligible on this nuclear scale. Gravity is one of the fundamental forces of nature, but it is most important on the astronomical scale. Planets, stars and galaxies are massive bodies which can attract one another over vast distances; it is only on the macroscopic scale that the gravitational attraction between protons could possibly be great enough to overcome their mutual electrostatic repulsion.

This leaves us with the electrostatic repulsion between protons, a force which tends to blow the nucleus apart. It therefore follows that another force exists – an attractive force, similar in magnitude to the electrostatic force and which holds the nucleus together. If such a force did not exist, all nuclei would be unstable and the Universe would be a very different place!

2.2 AN ADDITIONAL FORCE

Now that we have decided that an additional force must act between nucleons, what can we identify as the characteristics of this force, on the basis of our simple picture of the nucleus?

The force we are concerned with is called the **strong interaction** or **strong force**, and is distinct from the gravitational and electrostatic forces. It was investigated extensively by, amongst others, the Japanese physicist Hideki Yukawa; he identified the following characteristics of the strong force.

Fig 2.2 Hideki Yukawa, the Japanese physicist whose work in the 1930's led to an understanding of the nature of the strong nuclear force.

1. It is an attractive force between nucleons.

2. It is repulsive at very short range.

3. It does not extend beyond distances of a few femtometres.

4. It does not depend on the charge of the nucleons.

5. It is readily 'saturated' by surrounding nucleons.

We can understand these points from what we already know about the nucleus. We have already established that there must be some force holding the nucleus together. It must overcome the Coulomb repulsion between protons, and it must hold both protons and neutrons together. Hence we can safely say that the force must be attractive.

However, this is not enough; if the strong force was only attractive, it would pull the nucleons together into a vanishingly small region of space. We know that the nucleus appears to consist of distinct nucleons which retain their identity within the nucleus. The more nucleons, the bigger the nucleus. The centre-to-centre separation of the nucleons is approximately the distance r_0, which we discussed in Chapter 1. For this to be true, the strong force must be repulsive at very short range (less than about 1 fm).

There are several pieces of evidence to suggest that the strong force does not extend far outside the nucleus itself. A detailed analysis of the Rutherford scattering of alpha particles shows that the scattering is a result of the Coulomb repulsion between positive charges. However, if fast-moving protons are used, they can penetrate much closer to the nucleus where they experience the strong force, and a different scattering pattern is observed.

Another piece of evidence comes from the fact that the radioactive decay of nuclei is a random, spontaneous process. If one nucleus decays, this does not affect neighbouring nuclei, which are no more or less likely to decay as a result. Neighbouring nuclei in a solid are about 10^5 fm away, well beyond the range of the strong force.

Evidence from scattering experiments shows that the nuclear forces between protons and neutrons, between protons and protons, and between neutrons and neutrons are the same. (Of course, we have to take account of the Coulomb force between protons.)

Lastly, we must look at the idea of saturation of the strong force within a nucleus. The force extends only a very few femtometres from a nucleon. Within a nucleus, an individual nucleon is likely to be surrounded by

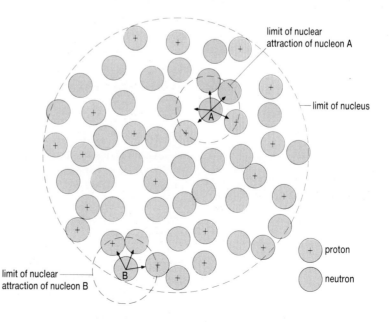

Fig 2.3 The stong force acts only between close neighbour nucleons in the nucleus.

Fig 2.4   The variation of the strong force between nucleons, as a function of their separation.

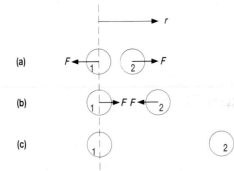

Fig 2.5   The force between two nucleons depends on their separation.

several other nucleons which are close enough to experience its attraction (see Fig. 2.3). However, other nucleons which are further away in the same nucleus do not experience its attraction; the force it exerts has been saturated by its nearest neighbours in the nucleus. Perhaps you can see how this is related to the fact that all nuclei have roughly the same density. We will look at this point in Section 2.3.

Force–distance graph

We can combine these very theoretical considerations into a simple picture of the strong nuclear force. Imagine two nucleons – two neutrons, for simplicity. When they are a long way apart, they exert no force on one another. When they are closer, they attract; closer still, and they start to repel. This is represented in Fig 2.4, a graph to show how the strong force depends upon the separation of the two nucleons.

This graph uses the sign convention that repulsive forces are positive, attractive forces are negative – see Fig 2.5. **(a)** When two nucleons are very close together, they repel one another. The repulsive force F exerted by nucleon 1 on nucleon 2 is in the direction of increasing separation r. F and r are thus both positive. **(b)** When the nucleons are further apart, they attract one another. F is now in the opposite direction to r, and so is negative. **(c)** When the nucleons are still further apart, there is no strong interaction between them.

Although we can represent the strong interaction between nucleons by a graph, we cannot provide a simple equation comparable to the equations (1) and (2) which describe the electrostatic and gravitational forces between particles. The strong interaction does not obey a simple inverse square law.

However, you should be able to see from the graph that, beyond a separation of about 1 fm, it decreases rapidly. At $r = 1$ fm, its value is about 8000 N; at $r = 2$ fm, its value is about 1400 N. This is faster than a $1/r^2$ variation.

QUESTIONS

Examine the graph in Fig 2.4.

2.5 Identify the region within which the force between two nucleons is attractive, and the region within which it is repulsive.

2.6 What is the greatest value of the attractive force between two nucleons?

2.7 At what small value of separation is the force between two nucleons zero? At this separation, what would be the electrostatic repulsion between two protons?

Fig 2.6   The van der Waals force between two krypton atoms.

A similar force–distance graph

You are probably familiar with the van der Waals force, which acts between neutral atoms or molecules. It is one of the forces which holds solids together. (You can read about the van der Waals force in most standard A-level physics textbooks. See Appendix A.)

The reason for referring to this force here is that it shares some of the characteristics of the nuclear strong force. Fig 2.6 shows the variation of this force with separation for two neutral atoms. You can see that there are regions of attractive and repulsive force, and the force becomes zero at large separations.

It is important to appreciate the differences in scales between these two graphs. To see the similarities, and the differences, between this graph and that of Fig 2.4, answer the questions which follow.

FORCES IN THE NUCLEUS

Examine the graph in Fig 2.6.

2.8 Identify the region within which the force between the two atoms is attractive, and the region within which it is repulsive.

2.9 What is the greatest value of the attractive force between the two atoms?

2.10 At what small value of separation is the force between the two atoms zero? At this separation, what would be the electrostatic repulsion between two protons?

Now that we have looked at the forces which bind the nucleus together, we can go on to look at the energy involved when nucleons come together to form a nucleus.

2.3 THE ENERGY OF NUCLEONS

Fig 2.7 Energy-separation graph for the strong nuclear interaction between two neutrons.

In order to understand the potential energy of nuclear particles, first consider a more familiar form of energy – gravitational potential energy. The Earth attracts you; it exerts a force on you which, according to our sign convention, is negative. If you climb to the top of a high building, you do work against gravity, and your gravitational potential energy increases. If you step off the top of the building, you rapidly return to a position of low energy, under the influence of the attractive force.

Now picture a neutron in a nucleus. It is held in by the attractive strong force. If you try to pull it away from the nucleus, you have to do work against the strong force. You are giving the neutron potential energy. Let it go, and it will be pulled back to a position of low energy.

The graph in Fig 2.7 shows the variation of energy W with separation r for two neutrons. It shows that the lowest energy state is when their separation is about 0.6 fm. For larger separations, the energy is greater and approaches zero. (Technically, W is defined to be zero for infinite separation.) For separations smaller than 0.6 fm, the energy again increases.

Energy and force

How can we relate this to what we know about the strong force? How does Fig 2.7 relate to Fig 2.4?

When a force F moves through a distance x in the direction of the force, it does work W given by:

$$W = F \times x$$

In calculus terms, we can write this in a more general form:

$$W = -\int F \mathrm{d}x$$

In other words, work done is represented by the area under the force–distance graph. Since integration is the reverse of differentiation, we can equally write the following:

$$F = \frac{-\mathrm{d}W}{\mathrm{d}x}$$

The force graph is thus found from the gradient of the energy graph. (The minus signs appear in these two equations for the following reason: when the two neutrons are separated, the nuclear force F does not do work; rather, you have to do work against it. So the work done by F is negative.)

The graphs of Figs 2.4 and 2.7 are shown again in Fig 2.8, one above the other, for ease of comparison. (Quantities are given in SI units.) You should identify the following points.

- Where the gradient of the energy graph is positive, the force is negative. Where the energy gradient is negative, the force is positive.

- Where the gradient of the energy graph is zero (at the minimum), the force is zero. This is the equilibrium separation of the two particles.

- The shaded area under the force graph represents the energy of the particles at the equilibrium separation.

2.11 Estimate this shaded area from the graph. What are the units of this quantity?

2.12 Compare it with the value of the energy minimum on the other graph.

Fig 2.8 Force–separation and energy–separation graphs for the strong interaction between two nucleons.

Nucleons in the nucleus

So far we have been considering the strong force between two nucleons. This force holds neutrons and protons together in the nucleus.

We also know about the Coulomb force which acts between protons. A proton outside the nucleus experiences Coulomb repulsion.

We can construct a simple picture of the energy of nucleons as follows. Since nucleons remain bound in a stable nucleus, they must be in a low energy state. Outside the nucleus, nucleons feel no force since the strong force is very short-range. Their energy is zero. Inside the nucleus, they experience attractive forces on all sides from the other nucleons. We say that the nucleons are trapped in a 'potential energy well'. This is illustrated, for neutrons, in Fig 2.9(a).

What this picture tells us is that the nucleus is a spherical region of space of radius r, in which the nucleons are in a low energy state.

For protons, the picture is complicated by the Coulomb repulsion. The graph of Fig 2.9(b) shows how the electrostatic potential energy of a proton varies near a nucleus. If you have studied the gravitational potential energy of a body near a solid sphere such as the earth, you should recognise the shape of this graph.

(a)

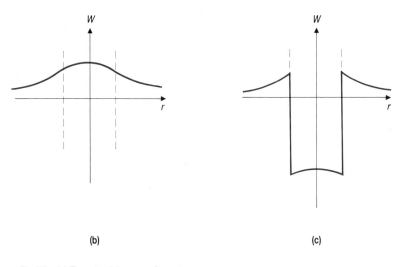

(b)

(c)

Fig 2.9 (a) The potential energy of a neutron.
 (b) The electrostatic potential energy of a proton.
 (c) The total potential energy of a proton, in the region of a nucleus.

As a proton approaches the nucleus it slows down, its kinetic energy decreases and its electrostatic potential energy increases. If it has enough energy initially it can penetrate the 'Coulomb barrier' and fall into the potential energy well. The graph shown in Fig 2.9(c) shows the potential energy–distance graph for a proton near a nucleus. This graph has been constructed by adding the graphs of Figs 2.9(a) and (b); it takes account of both the electrostatic and strong forces.

These graphs show the variation of potential energy along a line through the centre of the nucleus. We can picture the nucleus as being like a crater, perhaps the crater of a volcano, or a crater on the moon. The graph of Fig 2.9(c) is a cross-section through the crater; the nucleons of which the nucleus is composed are trapped within this crater.

Both graphs of Figs 2.9(a) and (c) are, of course, only approximations to the shape of the true potential well for a real nucleus. Here they are drawn as 'square wells'; we could construct better approximations by taking into account the detailed shape of the energy graph for two nucleons (Fig 2.7). In the questions below you are asked to consider an improved version of the potential well diagram for a proton.

We will make use of our simple picture of the energy of particles near a nucleus when we come to discuss the mechanisms of radioactive decay of the nucleus in Chapter 6.

QUESTIONS

Fig 2.10

The graph of Fig 2.10 shows an improved version of Fig 2.9(c). (Remember that the gradient of this graph tells you about the force on a proton.)

2.13 What does the flat bottom of the potential well suggest about the force on a proton within the nucleus? Why does the fact that the 'walls' of the well are sloping slightly make this a better model?

2.14 From the graph of Fig 2.9(c), estimate the least kinetic energy you would expect a proton outside the nucleus to require to overcome the Coulomb barrier.

FORCES IN THE NUCLEUS

2.4 MODELS OF THE NUCLEUS

The story so far....

In our discussion in these first two chapters we have been trying to build up a picture of the nucleus. We have looked at the constituent particles of the atom and the way in which they are distributed. We have looked at the approximate dimensions of the nucleus and the way in which this information has been obtained experimentally.

We have considered the forces at work in the nucleus, the two most important being the strong nuclear force acting between all nucleons, and the electrostatic Coulomb force, which acts only between protons.

In the following chapters, we are going to consider the behaviour of nuclei. Some nuclei are very stable, while others decay within a fraction of a second. Some small nuclei fuse together to make larger nuclei; some large nuclei split into smaller fragments. With the aid of modern accelerators, new reactions between nuclear particles have been discovered and studied.

In order to understand this unfamiliar world of nuclear reactions, it is necessary to have a picture in our minds of the nature of the nucleus. Current understanding of nuclear physics has been built up on the basis of many of the ideas which we have discussed in these two chapters.

We may think in terms of the forces between nucleons, or of the energy of nucleons. We described the nucleus as a potential well, or as a crater. We thought of the nucleus as being like a liquid drop – this is a model we will return to in more detail in Chapter 4. When we discuss the mechanism of gamma ray emission in Chapter 6, we will construct a very different picture of the nucleus.

All these ways of thinking about the nucleus are models. It is important to appreciate that each model of the nucleus is simply a way of describing some aspects of what we know of the way in which nuclei behave. We cannot say that a nucleus is exactly like a drop of liquid; rather, a nucleus has certain features in common with a liquid drop and, very importantly, this model allows us to make useful predictions about nuclear behaviour. Indeed, it may be that we choose to use two different models to explain different aspects of the nucleus, and that these different models appear to conflict. We must shrug our shoulders, accepting that any model must have its limitations, and make the most of the limited models which are available to us.

Models help us to understand our observations and to predict nuclear behaviour. Hopefully you have managed to build up a picture of the nucleus which will allow you to go on to understand some of the phenomena described in the following chapters. To see how well you have understood the story so far, try to answer the questions below.

QUESTIONS	
2.15	The nucleus may be probed using beams of energetic alpha particles, electrons, protons, neutrons and other particles. Explain, in terms of the forces acting between particles, why you might expect to get different information using different particles.
2.16	Sketch graphs to show how the strong nuclear force between two nucleons varies with their separation. Sketch a second graph to show how their potential energy varies with separation, and explain how the two graphs are related.
2.17	A nucleus is often described as a 'potential well'. Explain what this term means, and describe why there is a 'Coulomb barrier' to the penetration into the nucleus of positively charged particles.

SUMMARY

The gravitational force between nucleons is insignificant; the Coulomb repulsion between protons tends to blow the nucleus apart. A strong, short-range force exists which acts between all nucleons. This force is repulsive at very short distances, but attractive at slightly larger distances. It is this force which holds the nucleus together.

EXAMINATION QUESTIONS: Theme 1

T1.1

(a) Geiger and Marsden aimed alpha particles of mass 6.7×10^{-27} kg and speed 2.0×10^7 m s^{-1} at gold nuclei for which the number of protons is 79.

Calculate the closest distance of approach for a head-on collision, assuming that the gold nucleus remains stationary.

Charge of proton $= 1.6 \times 10^{-19}$ C.

Permittivity of free space, $\varepsilon_0 = 8.9 \times 10^{-12}$ F m^{-1}.

(b) Describe how electron diffraction is used in determining nuclear sizes. Explain why this method is used in preference to alpha particle scattering experiments.

(JMB 1985)

T1.2

In an alpha particle scattering experiment, an alpha particle and a gold nucleus (in a piece of gold foil) collide head-on and the alpha particle rebounds.

(a) Using the data below, write down a numerical expression for the electrostatic force of repulsion, F, acting on the alpha particle at the instant of collision when the distance between the alpha particle and the gold nucleus is s.

For the alpha particle $A = 4$, $Z = 2$; for the gold atom $A = 197$, $Z = 79$.

Electronic charge, $e = 1.6 \times 10^{-19}$ C.

Permittivity of vacuum, $\varepsilon_0 = 8.85 \times 10^{-12}$ Fm^{-1}.

(b) The electric potential energy of the alpha particle and the gold nucleus at the point of impact is $F s$.

(i) If the initial kinetic energy of the alpha particle is 1.8 MeV, calculate a value for s. Assume that the gold nucleus has no kinetic energy initially. (1 eV $= 1.6 \times 10^{-19}$ J.)

(ii) What indication does the value of s give about the size of the two particles involved?

(ULSEB 1988)

T1.3

(a) Explain why electron diffraction is used to determine nuclear size.

(i) Estimate the momentum the electrons must be given.

(ii) Name an apparatus capable of producing electrons having appropriate speeds.

Planck's constant $h = 6.6 \times 10^{-34}$ J s.

(b) What are the main characteristics of the interaction between nucleons in the nucleus? Show how they account for the relationship $R = r_0 A^{1/3}$, where:
R = nuclear radius
A = atomic mass number
r_0 = a constant.

(JMB 1981)

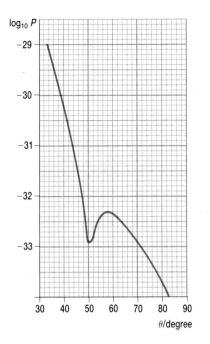

T1.4

The graph shows the results of electron scattering experiments for ^{12}C, in which electrons of energy 420 MeV are used. P is a measure of the number of electrons scattered at angle θ to the incident beam.

(a) Show that the de Broglie wavelength, λ, of the incident electrons is 2.96×10^{-15} m, assuming that the relation energy/momentum = c, the speed of light, applies at this energy.

The charge of an electron = -1.60×10^{-19} C

Speed of light = 3.00×10^8 m s^{-1}.

The Planck constant = 6.63×10^{-34} J s.

(b) Give two reasons why high energy electrons are used in determining nuclear size. What information do such measurements give about nuclear density and the average separation of particles in the nucleus?

(c) In diffraction theory, the first minimum of the scattered intensity produced by a spherical object of radius R occurs at an angle θ to the direct beam, where θ is given by

$$\sin \theta = \frac{0.61 \lambda}{R}$$

Use the experimental data to calculate the radius of the ^{12}C nucleus.

Use your result to calculate the radius of the ^{16}O nucleus.

(JMB 1984)

T1.5

(a) The table opposite gives a series of values of R, the radius of the nucleus, as determined from electron scattering experiments, for nuclides of atomic mass number A. Plot a suitable straight line graph to show the relationship between R and A. Explain how your graph shows that the density of nuclear matter is approximately constant. Use your graph to calculate this density, assuming that the mass of a nucleus is 1.7×10^{-27} kg.

Element	A	$R/10^{-15}$ m
silicon	28	3.70
vanadium	51	4.50
cobalt	59	4.82
strontium	88	5.34
indium	115	5.80
antimony	122	5.97
bismuth	209	7.13

(b) What can be deduced about the forces between nucleons from the fact that the nuclear density is constant?

(JMB 1982)

T1.6

(a) Draw sketch graphs of **either** *potential energy*, **or** *force*, as a function of separation to show the main qualitative features of the interaction between (i) two neutrons, and (ii) two protons, at separations comparable to the sizes of nuclei. Indicate approximate scales on the axes of your graphs.

(b) Explain how the nature of the nuclear interaction accounts for the relationship between nuclear mass number and nuclear radius.

(c) Estimate the radius of a nucleus of ^{64}Zn.

(JMB 1980)

Theme 2

MASS AND ENERGY, STABILITY AND DECAY

The radioactive nature of some materials, particularly those containing the element uranium, was first shown by Henri Becquerel at the end of the nineteenth century. His work was followed up by the Polish born scientist Marie Curie. She and her husband Pierre devised a technique for determining the degree of radioactivity of different substances. She showed that the activity of uranium compounds depended on the amount of uranium they contained. It did not depend on whether the uranium was chemically combined with other elements, or on its physical state – solid, liquid or gas. This vital observation showed that radioactivity was an atomic phenomenon; the Curies went on to identify other radioactive elements, including thorium, polonium and radium.

We can now make radioactive sources which may be used conveniently in school and college laboratories to provide alpha, beta and gamma radiations. You have probably seen and carried out experiments with such sources yourself. The nature of these radiations, and their origins within the atomic nucleus, have been explored throughout the century since the work of Becquerel and the Curies. In studying this Theme you will develop a picture of some models which describe and explain the phenomenon of radioactivity.

Marie and Pierre Curie in their laboratory, 1896.

A school laboratory today.

Chapter 3

STABLE AND UNSTABLE NUCLEI

This chapter shows how you can discover information about the properties of nuclides, both stable and unstable. The skill to find out about nuclides will prove useful in subsequent chapters.

LEARNING OBJECTIVES

After studying this chapter you should be able to:

1. extract information about nuclear masses, decay modes, isotopic abundances and half-lives from tables and charts of nuclides;

2. show how nuclides are related through decay chains.

3.1 FINDING OUT ABOUT NUCLIDES

More than 100 different elements have been observed; some of these have been artificially created. Of these elements, more than 2000 different nuclides, most of them artificial, are now known.

You will remember that nuclides differ in their numbers of protons and neutrons. If two nuclei have the same values of Z and N, then they are identical nuclides.

This may seem to be a bewildering array of different nuclear species. By comparison, chemistry might seem simpler than nuclear physics, since it only has to deal with the 100 or so elements.

In this section we will look at some of the information available about nuclides and try to make sense of it.

Tables of isotopes

One way in which information about nuclides is presented is as a table of isotopes. In the assignment which follows, you are asked to explore such a table.

ASSIGNMENT

Find a table of isotopes. Some suitable books of scientific data are referred to in Appendix A.

Look at the headings of the different columns of the table. You should find that the table tells you values of Z and A. It will probably also tell you some or all of the following: abundance of stable nuclides; half-life and mode of decay of unstable nuclides; mass in u; energies of emitted particles or photons.

The following questions will help you to work your way round such a table.

3.1 Which are the two stable isotopes of chlorine (Cl, $Z = 17$)? What are their abundances?

3.2 Radiocarbon dating depends on the decay of $^{14}_{6}C$. What is the half-life of this decay? What particle (α or β) is emitted?

3.3 How many stable isotopes of lead (Pb, $Z = 82$) are there?

3.4 A common source of γ rays in school laboratories is the isotope $^{60}_{27}$Co. What are the energies in MeV of γ photons from such a source?

3.5 What is the stable nuclide with the greatest mass?

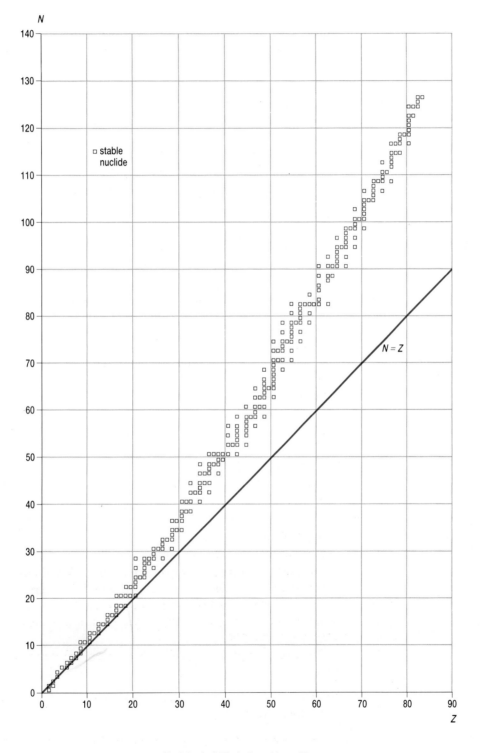

Fig 3.1 An *N-Z* plot for stable nuclides.

Stable nuclei

A table of isotopes is not the clearest way to present information if we are to find patterns in the properties of all these hundreds of different nuclides. In order to make sense of all this data, let's start by concentrating on stable nuclei.

Which nuclei are stable? We can present this information visually, in the form of a graph of N against Z. Such a graph is shown in Fig 3.1. Since each nuclide has a unique combination of values of N and Z, it follows that each point on this graph represents a single stable nuclide.

We can find out some empirical facts about the stable nuclei by inspecting this graph. Look for the points which correspond to the smallest stable nucleus, ^1_1H, and the largest, $^{209}_{83}Bi$. These are the extreme range of the stable nuclides.

The graph shows a straight line corresponding to $N = Z$. For the lightest nuclei, the stable isotopes fall close to this line. For higher values of Z, you should be able to see that $N > Z$. Massive stable nuclei have more neutrons than protons. Use this graph to answer the following questions.

QUESTIONS	
	3.6 Above what approximate value of Z does the 'line of stability' deviate from the relationship $N = Z$?
	3.7 What is the approximate value of the ratio $N : Z$ for the most massive stable nuclei?
	3.8 Why do you think the most massive nuclei, which have a large number of protons, require a higher proportion of neutrons to remain stable? (Hint: think about the forces involved.)
	3.9 All the isotopes of an element (**p** for proton) have the same value of Z. How do they appear on this graph? Isotones (**n** for neutron) have the same value of N. How do they appear on this graph?

Empirical rules for stability

From these observations, we can write two rules for nuclear stability:

$$N \approx Z \text{ for } Z < 20 \qquad\qquad N > Z \text{ for } Z > 20$$

These are empirical rules; that is, we have arrived at them from an analysis of the available data, presented in Fig 3.1. We do not have any theoretical explanation of them.

You may have observed another pattern in the data of the graph. Some values of Z seem to appear more frequently than others. Some values of N also appear more frequently. For example, five nuclides have $N = 20$; only one has $N = 19$, and only one one has $N = 21$.

Perhaps you can see from the graph that even values of N seem to be the most 'popular'; similarly, even values of Z appear most frequently. Table 3.1 shows an analysis of the stable nuclei, divided up according to whether they have even or odd values of N and Z.

Table 3.1
Numbers of stable nuclides.

Z	N	Number
even	even	160
even	odd	56
odd	even	52
odd	odd	4
		total = 272

QUESTIONS	
	3.10 From Table 3.1, which is the most common combination of values of N and Z?
	3.11 Only four stable nuclides are 'odd–odd'; that is, they have odd numbers of both protons and neutrons. All of these have nucleon numbers A less than 20. Use a table of isotopes to identify them.
	3.12 What is the lightest 'even-even' stable nuclide? What is another name for this particle?

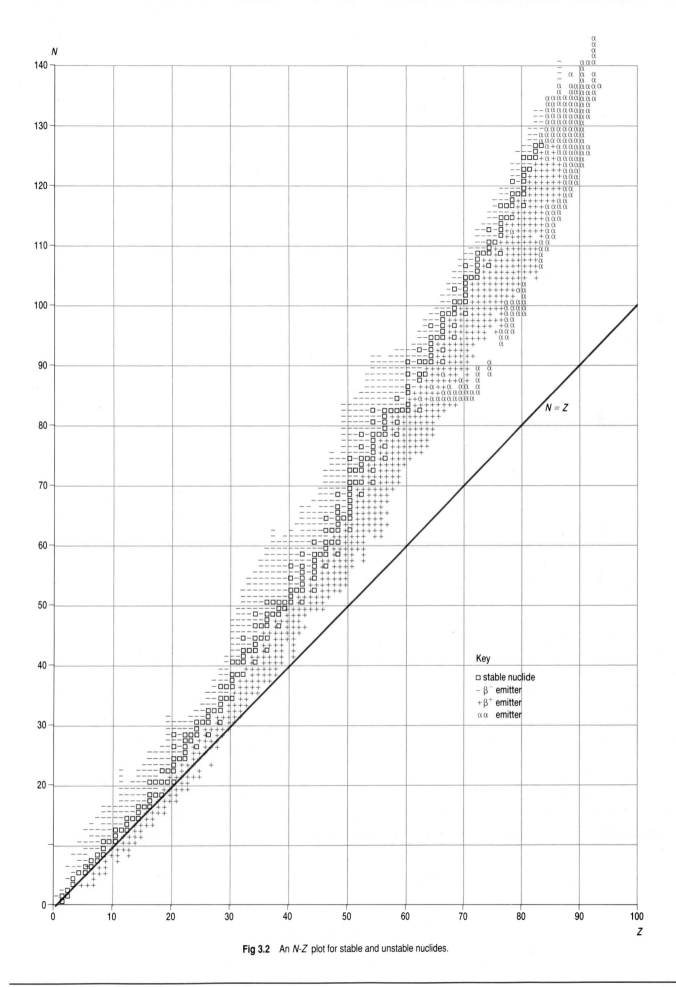

Fig 3.2 An *N-Z* plot for stable and unstable nuclides.

In this section we have looked at the stable nuclei and tried to find a pattern which governs their stability. In the next section, we will look at the unstable nuclei (which are radioactive) and try to find some more patterns to help us understand the behaviour of the atomic nucleus.

3.2 UNSTABLE NUCLEI

You have already looked at a table of isotopes, which gives information about which nuclides are stable and which are radioactive. It also tells you what type of radioactive decay each unstable nuclide exhibits.

Such a table is not a very clear way of displaying this information if we want to look for a pattern amongst the hundreds of different nuclides. An alternative way is to construct a chart of the nuclides. This is usually an N–Z plot similar to Fig 3.1, with all the nuclides shown. Such a chart shows whether a nuclide decays by alpha or beta emission, or in some other way.

For the following assignment, you will need to examine such a chart. Fig 3.2 shows an N–Z plot with some radioactive nuclides, as well as the stable ones. (A much bigger graph is needed to show all the known nuclides in detail.)

ASSIGNMENT

Find a suitable chart of the nuclides (or use Fig 3.2). Make sure you know which quantities are used for the axes. Identify the line of stability. Answer the questions which follow. (In these questions, and the discussion after them, I have assumed that your chart shows N on the y-axis, and Z on the x-axis.)

3.13 How are the α emitters indicated on the chart? Where do they appear on the chart, relative to the line of stability?

3.14 There are two types of β emitters. Look for the β^- emitters. (These are the familiar sources of β radiation, such as the ^{90}Sr which you may have used in the laboratory. You should recall that a β particle is an electron.) Where are these unstable nuclides concentrated on the chart?
A second type of β emitter, denoted β^+, is also shown on a nuclide chart. Where can these be found, relative to the line of stability?

Empirical rules for α and β decay

From these observations of the distributions of unstable nuclides on the chart we can invent some empirical rules which describe which nuclides decay by α, β^- or β^+ emission. We can think of nuclides which lie above the line of stability as having too many neutrons to be stable; they are described as 'neutron-rich'. They decay by β^- emission. Nuclides which lie below the line have too many protons to be stable; they are described as 'proton-rich'. These nuclides decay by β^+ or α emission. It seems that, in these decay processes, nuclei are adjusting the proportions of protons and neutrons to achieve stability. If we look in more detail at these mechanisms of decay we can see more clearly how this happens.

α decay

An α particle consists of two protons and two neutrons. When a nucleus emits an α particle the values of both Z and N decrease by two. Why does this make the nucleus more stable? Fig 3.3 shows an enlarged portion of the N–Z plot of nuclides. An α emitting nuclide, $^{212}_{84}$Po is shown; this nuclide decays to $^{208}_{82}$Pb, which is stable. This decay is indicated by the arrow on the figure. Alpha decay thus appears as a diagonal arrow on the figure, bringing the unstable nuclide closer to the line of stability.

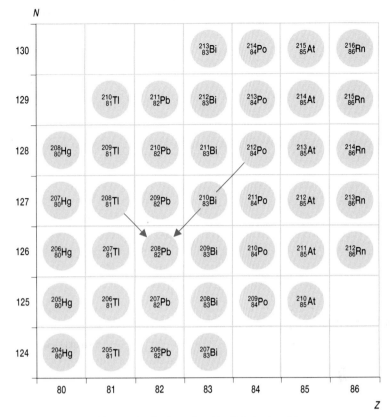

Fig 3.3 Details of a portion of the N-Z plot of nuclides.

β⁻ decay

A β⁻ particle is an electron, which is negatively charged. When a nucleus emits a β⁻ particle it becomes more positively charged. Its mass is barely affected. The value of Z increases by one, while the value of N decreases by one. (It is as if a neutron had changed to a proton, emitting an electron in the process.) In Fig 3.3, you can see a β⁻ emitter, $^{208}_{81}$Tl. This decays to $^{208}_{82}$Pb. This decay is represented by an arrow, shorter than that for the α decay (since N and Z change by only one). Again, by decaying, the unstable nuclide is brought closer to the line of stability.

β⁺ decay

This is a form of radioactive decay which may well be unfamiliar to you. A β⁺ particle is emitted by an unstable, proton-rich nucleus. You may have guessed from the symbols used that a β⁺ particle is a positively charged electron. It is sometimes known as a positron, or antielectron. This is an example of an antiparticle, a particle of antimatter. Later you will become acquainted with more antiparticles. For the moment it is enough for you to picture a β⁺ particle as being like an electron, but with a charge of $+e$.

γ emission

We will discuss γ emission in detail in Chapters 5 and 6. For the moment it is enough to know that α and β decay are often accompanied by the emission of a γ photon. Since this is an electromagnetic photon, which has no charge, it does not affect either Z or N, and so this process is not relevant to our present discussion.

If you have understood how N and Z change during α and β⁻ decay, and how these are represented in Fig 3.3, you should be able to work out for yourself the details of β⁺ decay in the following questions.

3.15 When a nucleus emits a β⁺ particle, the nucleus becomes less positively charged. How do N and Z change?

3.16 Which nuclide shown in Fig 3.3 would you predict to decay by β⁺ emission to give the stable nuclide $^{208}_{82}\text{Pb}$? Check your prediction against a table or chart of nuclides.

3.17 How would you draw an arrow on the figure to show this decay?

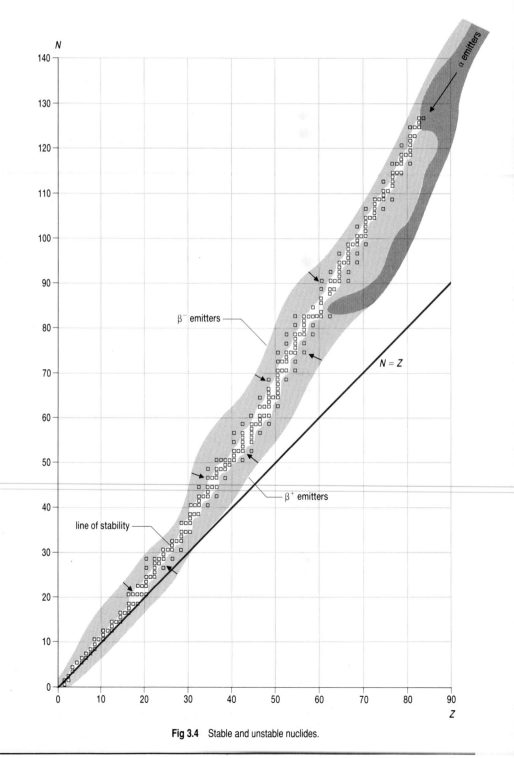

Fig 3.4 Stable and unstable nuclides.

STABLE AND UNSTABLE NUCLEI

Fig 3.4 summarises the information we have gathered about α, β⁻ and β⁺ emitters. It shows where they typically appear on the *N–Z* plot, and how they decay towards the line of stability.

3.3 DECAY CHAINS

When a radioactive nuclide decays it does not necessarily become stable; it may simply become a different unstable nuclide. The initial nuclide is called the **parent**; the product is a **daughter nuclide**. An unstable daughter is the parent of a further daughter, and so on. The decay of one nuclide to form another, which decays to form another, and another, is called a **decay chain** or **radioactive series**.

Fig 3.5 Five members of a decay chain.

A part of such a sequence is shown in Fig 3.5. (This sequence includes the nuclide $^{234}_{91}$Pa, an isotope of protactinium whose half-life you may well have measured. It is formed by the gradual decay of $^{238}_{92}$U in solution; it is then extracted into an organic solution in a sealed bottle, and the half-life – about 72 s – determined.)

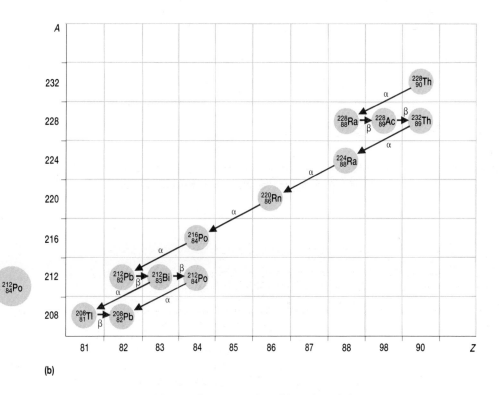

(a)

(b)

Fig 3.6 Two representations of the 4n decay chain.

Decay chains are represented in different ways. Fig 3.6 on the previous page shows two of these which you may encounter. In Fig 3.6(a) the nuclides of a chain are shown in sequence, with the mode of decay (α or β emission) indicated. In Fig 3.6(b) the same chain is represented as part of a plot of A against Z. (It could equally well have been plotted as N against Z.) In this case, α emission is represented by a diagonal line, and β emission by a shorter horizontal line. Both of these representations show an interesting feature: the chain branches at $^{212}_{83}\mathrm{Bi}$. This nuclide can decay in two ways; by emitting an α particle to become the β emitter $^{208}_{81}\mathrm{Tl}$, or by emitting a β particle to become the α emitter $^{212}_{84}\mathrm{Po}$. In either case, the final daughter product is the stable nuclide $^{208}_{82}\mathrm{Pb}$.

ASSIGNMENT

3.18 Look at the nuclides represented in the decay chain of Fig 3.6. Why do you think this is referred to as the '4n' series?

3.19 Look at the partial decay chain in Fig 3.5. How do you think we might refer to this series?

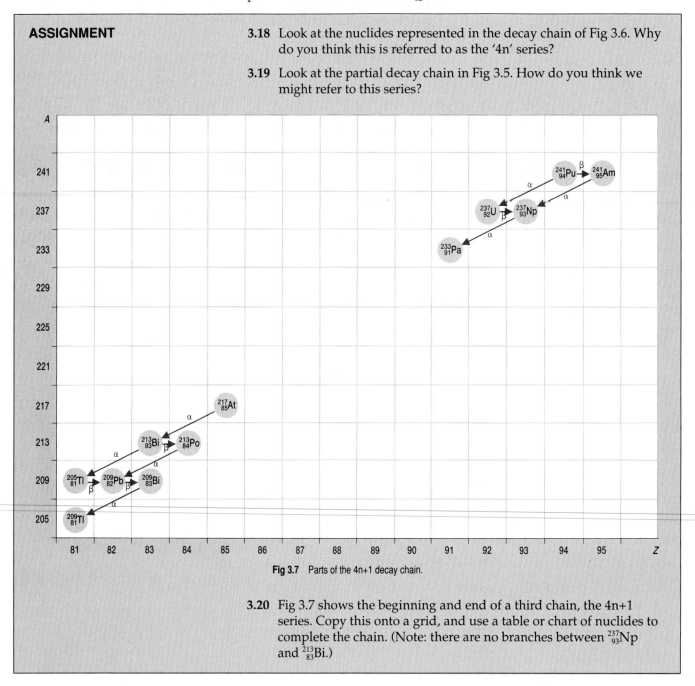

Fig 3.7 Parts of the 4n+1 decay chain.

3.20 Fig 3.7 shows the beginning and end of a third chain, the 4n+1 series. Copy this onto a grid, and use a table or chart of nuclides to complete the chain. (Note: there are no branches between $^{237}_{93}\mathrm{Np}$ and $^{213}_{83}\mathrm{Bi}$.)

You have now looked at three ways in which information about nuclides is presented: tables and charts of nuclides, and decay chains. You may also have looked at the computer software *Nuclides database*. All of these are useful in their own ways, and hopefully you will be able to refer to them as

STABLE AND UNSTABLE NUCLEI

appropriate in the next chapters. To see how well you understand these ideas, try the following questions.

QUESTIONS

A fourth decay chain, not referred to above, is known. It is not found in nature, although its members are found in nuclear reactors. This is the 4n+3 series.

3.21 The first nuclide in this series is $^{235}_{92}$U. What particle does this nuclide emit when it decays? What is the daughter product of this decay? What is the half-life of this decay?

3.22 This decay series ends with a stable isotope of lead, $_{82}$Pb. Deduce the value A for this nuclide. (Start by looking up the stable nuclides of lead.)

3.23 How many α and how many β decays must there be between the two ends of this series?

3.24 The decay of the initial nuclide, $^{235}_{92}$U, has the longest half-life of any nuclide in this series. All other decays are considerably faster. What has this to do with the fact that this series is not observed in nature? Might it ever have been?

SUMMARY

Information about nuclides is displayed in different ways: as tables or charts, or as decay chains. You can find out about isotopic masses, abundances, stabilities, modes of decay and half-lives. Examination of this evidence shows that stable nuclides have $Z \approx N$ (for small Z), and $N > Z$ (for $Z > 20$). Modes of radioactive decay depend upon the proton:neutron ratio. Light, neutron-rich nuclei are β^- emitters; light, proton-rich nuclei are β^+ emitters; massive, proton-rich nuclei are α emitters.

Chapter 4

THE MASS–ENERGY BALANCE

We have now looked at the forces which hold the nucleus together and seen something of the pattern of instability amongst nuclides. In working through this chapter you will be able to develop a more detailed understanding of the factors governing the stability of the different nuclides.

LEARNING OBJECTIVES

After studying this chapter you should be able to:

1. calculate values of mass defect and binding energy for different nuclides;

2. use Einstein's equation $E = mc^2$ to relate mass defect and binding energy;

3. describe the variation of binding energy per nucleon with nucleon number A;

4. explain the significance of the terms in the semi-empirical binding energy equation.

4.1 NUCLEAR MASSES

The table of nuclides which you inspected in the previous chapter probably included values of masses. These are usually atomic masses; that is, they are not nuclear masses but include the mass of the electrons which, together with the nucleus, make up the atom.

Fig 4.1 The principle of operation of one form of mass spectrometer.

Values of mass are quoted like this because of the way in which they are measured. It is not possible to experiment with nuclei free of electrons. Instead, a **mass spectrometer** is used; a beam of ions is deflected by electric and magnetic fields. The amount of deflection depends on the charge-to-mass ratio of the ions. (This is similar to the familiar experiment to measure e/m for the electron.) Knowing the charge of the ions in question, we can find their mass.

The principle of operation of one form of mass spectrometer is illustrated in Fig 4.1.

To calculate the mass of a particular nuclide it is only necessary to subtract the mass of the appropriate number of electrons. The mass of one electron is 0.000 549 u, to 6 decimal places.

Atomic masses have been determined experimentally to a high degree of accuracy, in some cases to better than one part in 10^8. It may seem surprising to work to such a high degree of precision. We do not often need to work with numbers which are accurate to eight significant figures. However, in the rest of this chapter, you will come to see why this is necessary in nuclear physics.

The questions which follow involve some simple calculations using values of atomic and nuclear masses.

QUESTIONS

In these questions, take $m_e = 0.000\ 549$ u.

4.1 The atomic mass of $^{16}_{8}\text{O}$ is 15.994 915 u. Calculate its nuclear mass.

4.2 You should know the atomic mass of $^{12}_{6}\text{C}$, since this is used to define the unit u. What is the nuclear mass of $^{12}_{6}\text{C}$? What fraction of the atomic mass is due to the electrons?

4.3 The nuclear mass of $^{20}_{10}\text{Ne}$ is 19.986 950 u. What is its atomic mass?

4.2 NUCLEAR MASS DEFECT

Significant figures

A striking feature of nuclear masses is that they are all nearly, but not quite, integers. The neutron and the proton have masses close to 1u. When these are combined together to give a nucleus, how does the nuclear mass compare with the sum of the masses of its constituent nucleons?

We can investigate this most simply by looking at an example, the nuclide $^{4}_{2}\text{He}$. This is made up of two protons and two neutrons.

We start by picturing these nucleons as four separate particles. Their total mass is given by:

$$M(2p + 2n) = 2m_p + 2m_n$$
$$= (2 \times 1.007\ 276) + (2 \times 1.008\ 665)$$
$$= 4.031\ 882 \text{ u}$$

This is the mass of the four separate nucleons. Now we picture the same nucleons bound together as a nucleus, $^{4}_{2}\text{He}$. The mass of this nuclide (taking care to subtract the mass of its two electrons) is given by:

$$M(^{4}_{2}\text{He}) = 4.001\ 506 \text{ u}$$

Compare these two values of mass. Here we have one of the most striking results of nuclear physics. The mass of the nucleus is less than the sum of the masses of its constituent parts. It appears that, as the nucleons come together to form the nucleus, their total mass decreases. In the case of $^{4}_{2}\text{He}$, there is a decrease in mass of 4.031 882 – 4.001 506 = 0.030 376 u . We say that there is a mass defect of 0.030 376 u .

The symbol for mass defect is ΔM. We can summarise the way to calculate ΔM as:

$$\Delta M = M(\text{separate nucleons}) - M(\text{nuclide})$$

Fig 4.2 summarises the idea that the separate nucleons have more mass than the nucleus which they become.

4.3 MASS AS ENERGY

It turns out that, for all stable nuclei, there is a mass defect. It appears that when neutrons and protons come together to form a nucleus, the total mass of the system decreases. How can this be?

In order to understand why this is so, we need to recall some of the ideas developed in Chapter 2. In particular, remember that when a nucleon is in a nucleus, it is in a lower energy state than when it is outside the nucleus. (You may have to look back to Fig 2.7.) In order to pull a nucleon out of the nucleus we have to give it energy. We do work against the attractive forces which are holding it in the nucleus.

Now we see that, in pulling a nucleon out of the nucleus, we are increasing not only its energy but also its mass. If you could pull a nucleus apart into its constituent nucleons, you would have to do work. You lose energy, and the energy you lose goes to increasing the mass of the nucleons.

At first, this is a rather strange idea. We are saying that the energy which you put into separating the nucleons is transformed into mass. This is not something which we experience in our everyday lives. It appears to conflict with the Principle of Conservation of Energy. How can we understand this better?

Einstein's equation

If we could dismantle a nucleus we would be increasing the potential energy of the nucleons. We have seen that the nucleons have greater mass when separated than when they are bound within the nucleus. We can only conclude that the increase in potential energy of the nucleons manifests itself as an increase in the masses of the nucleons. In other words, the energy which you put in to dismantle the nucleus has become the increase in mass of the nucleons.

This is an unfamiliar effect which we do not expect to observe. Energy converted into mass? This must mean that mass is a form of energy. In measuring the mass of a nucleus we are determining its potential energy.

Mass and energy are two ways of describing the same thing; it is rather like having two scales of temperature, Celsius and Kelvin. Both measure the same thing and we can convert from one to the other.

This is very important on the nuclear scale. Measurements of mass can be converted to energy, and energy to mass. The conversion is governed by an equation – perhaps the most famous equation in science – first deduced by Einstein:

$$E = mc^2$$

Fig 4.3 (a) Albert Einstein playing the part of the mad scientist, and (b) his notebook.

This tells us that energy E is proportional to mass m, and the conversion factor is c^2, the square of the speed of light in free space.

QUESTIONS

4.7 Calculate the energy equivalent to a mass of 1 kg.

4.8 Calculate the mass equivalent to an energy of 1 J.

The answers to these two questions show that, on a human scale, a small amount of mass is equivalent to a very large amount of energy.

4.9 In section 1.6, you showed that 1u is approximately equal to 1.661×10^{-27} kg. Use Einstein's equation to convert 1u into energy units: **(a)** J, and **(b)** MeV.

How $E = mc^2$ affects *you*

Einstein's equation is essential to our understanding of the nucleus. It is not of much importance to us in understanding ordinary events in our everyday lives. The assignment which follows shows why.

ASSIGNMENT

You are a student of mass 50 kg. According to Einstein's equation, if you increase your energy, your mass also increases. There are several ways in which you can increase your energy:
(a) you could run at 10 m s^{-1};
(b) you could climb to the top of a 1000 m high mountain;
(c) you could eat a sandwich which gives you 1 MJ of energy.

4.10 Calculate the increase in your energy and mass in each case. By what fraction does your mass increase?

It may surprise you to find that, by climbing to the top of a mountain, you have increased your mass by increasing your gravitational potential energy.

But this is precisely what happens when a nucleon is separated from a nucleus; its potential energy increases as work is done against the inter-nucleon forces, and this appears as an increase in mass. In the case of climbing a mountain, and the other examples above, your increase in mass is too small to detect.

4.4 BINDING ENERGY

Putting the nucleus together

We can picture making a nucleus as follows: Start with some protons and neutrons, all far apart from one another. Allow them to come together to form a nucleus, under the attractive influence of the strong nuclear force. They are in a lower energy state when they have formed the nucleus than when they were apart; i.e. they have lost energy. This energy is given out as a γ photon, or in some other way. This process is illustrated in Fig 4.4.

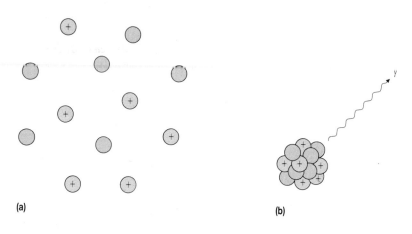

(a)   (b)

Fig 4.4 When separate nucleons (a) come together to form a nucleus (b) their energy decreases.

In order to pull a nucleus apart, energy must be supplied. You have to do work. The amount of energy required to pull a nucleus apart into its separate nucleons is called its **binding energy**, ΔE.

Binding energy and mass defect

To dismantle a nucleus you have to put in an amount of energy ΔE. This energy ends up in the form of mass; this is the increase of mass, called the mass defect ΔM, which we discussed in Section 4.2. It follows that ΔE and ΔM are related by Einstein's equation:

$$\Delta E = \Delta Mc^2$$

Since you know how to calculate mass defects, you can now also calculate binding energies for different nuclides. Since ΔM is usually given in u, and ΔE is most appropriately given in MeV, it is necessary to use the conversion factor between u and MeV. You calculated this factor in answer to question 4.9. The value we will use in calculations is:

$$1 \text{ u} = 931.3 \text{ MeV}$$

4.11 In Section 4.2 we found that the mass defect of $^4_2$He is 0.030 376 u. Calculate **(a)** the binding energy for this nuclide, in MeV, and **(b)** the binding energy per nucleon.

4.12 Use your answers to questions 4.4 and 4.5 to find the binding energy per nucleon for $^3_2$He and for $^{16}_8$O.

4.13 Which of these three nuclides do you think is the most tightly bound together?

Removing a nucleon

In a similar way, we can calculate the energy needed to remove a single nucleon from a nucleus. For example, if we remove a neutron from a nucleus of $^{13}_6$C, we get a nucleus of $^{12}_6$C. How much energy would this require? We need to calculate the increase in mass after the neutron has been removed.

$$
\begin{aligned}
\text{mass before} &= M(^{13}_6\text{C}) & &= 13.000\,060\ \text{u} \\
\text{mass after} &= M(^{12}_6\text{C} + \text{n}) &&= 11.996\,706 + 1.008\,665\ \text{u} \\
& && = 13.005\,371\ \text{u} \\
\text{increase in mass} & && = 0.005\,311 \times 931.3\ \text{MeV} \\
\text{energy required} & && = 4.95\ \text{MeV}
\end{aligned}
$$

QUESTION

4.14 If a neutron is removed from the nuclide $^4_2$He, it becomes $^3_2$He. Use the information in Section 4.2 about masses of these nuclides to find how much energy is needed to do this.

Binding energy per nucleon

A small nucleus has only a small value of binding energy; a large nucleus has a large value. In order to compare values for different nuclides, it is

Fig 4.5 The binding energy per nucleon for a number of stable nuclides.

usual to calculate the binding energy per nucleon. (You have already calculated this quantity in questions 4.11 and 4.12.) This gives us some idea of how strongly bound the nucleons are within the nucleus.

$$\text{binding energy per nucleon} = \Delta E / A$$

The results of determinations of binding energy per nucleon for many nuclides are shown in Fig 4.5, as a function of A. You can see that, for most nuclides, this quantity is between 7 and 9 MeV per nucleon. Some lighter nuclides have lower binding energies than this.

The most stable nuclides are those near the top of the graph, with $A \approx 60$. (It may seem that the graph is upside-down, but it is traditionally shown this way. Remember that binding energy is energy which has been lost in forming the nucleus. The greater the binding energy, the lower the energy of the nucleus, and thus the more tightly bound and stable it is.)

Most of the nuclides fall close to the smooth curve shown. You may notice that ^4_2He, the alpha particle, is in a stable position, some distance above the curve.

In the next section we will look at the factors which contribute to nuclear stability, and see how our understanding of these factors can explain the curve on the graph of Fig 4.5.

4.5 THE BINDING ENERGY FORMULA

The formula for the binding energy of nuclei was first derived by thinking about the nucleus as being like a droplet of liquid. (We first came across this model of the nucleus in Chapter 2.)

Refer back to Fig 2.3 for a picture of the interactions between nucleons within the nucleus. It is important for our discussion to appreciate the different situations of the nucleons labelled A and B in this figure; A is within the volume of the nucleus, surrounded on all sides by other nucleons, while B is on the surface, with fewer close neighbours.

We will start by looking at the formula for binding energy and then we will break it up into separate terms to see what they mean individually.

$$\text{binding energy } \Delta E = aA - bA^{2/3} - cZ^2A^{-1/3} - \frac{d(N - Z)^2}{A}$$

This expression is known as the semi-empirical binding energy equation. By breaking the equation down into separate terms, and considering where they come from, you will see that it is not quite as complicated as it looks at first sight. The coefficients a, b, c and d are constants found from experimental data. They are adjusted to make the equation fit the graph of Fig 4.5. Values of the coefficients are shown in Table 4.1.

Terms of the equation

We will consider the terms of the equation individually, and then see how the equation relates to the liquid drop model of the nucleus. Remember that, in calculating the binding energy, we are calculating the amount of energy required to separate the nucleus into its constituent nucleons, by breaking all the bonds between neighbouring nucleons.

aA: the volume term

The number of bonds depends directly on the number of nucleons A. (To remove a nucleon, such as A in Fig 2.1, requires a certain amount of energy. The more nucleons there are, the more energy is needed.)

−bA²/³: the surface term

The volume term overestimates the binding energy, since nucleons such as B in Fig 2.3 which are near the surface have fewer neighbours, and hence are less tightly held within the nucleus. This term is a correction factor.

Table 4.1

Values of the coefficients of the semi-empirical binding energy formula.

Coefficient	Value/MeV
a	14.0
b	13.0
c	0.58
d	19.3

THE MASS–ENERGY BALANCE

Since the radius of the nucleus is proportional to $A^{1/3}$, the surface area is proportional to $A^{2/3}$.

$-cZ^2A^{-1/3}$: the Coulomb term

The Z protons within the nucleus attract one another via the strong nuclear force. They also repel each other electrostatically. They have electrostatic potential energy; this is proportional to (charge^2/radius), in other words to $Z^2/A^{1/3}$.

$-d(N-Z)^2/A$: the symmetry term

We saw in Section 3.1 that nuclides with $N = Z$ tend to be the most stable. This term has the effect that, the bigger the difference between N and Z, the greater the reduction in the binding energy. This makes the nucleus less stable. For a nucleus with $N = Z$, this term is zero.

The liquid drop model

How do these terms relate to the picture of a nucleus as a drop of liquid? If we give energy to a liquid drop, what happens?

Firstly, bonds between neighbouring molecules begin to break as the temperature of the liquid starts to rise. For the drop to break up into individual molecules, it must boil. The heat energy required to do this is the latent heat. The volume term in the binding energy equation is often called the latent heat term. Secondly, molecules on the surface of the drop are less tightly bound. They are bonded to fewer neighbours, and can escape relatively easily (evaporate). (The other terms in the equation do not have direct equivalents in the liquid drop model.)

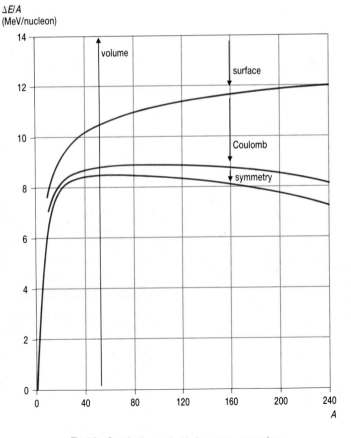

Fig 4.6 Contributions to the binding energy per nucleon.

Binding energy graphs

Fig 4.6 on the previous page shows how these terms contribute to the curve of Fig 4.5. The graphs show how the binding energy per nucleon varies with A for each term in the equation. (The terms have been divided by A.)

Inspect the binding energy formula on page 48 and the graphs of Fig. 4.6, then answer the following questions:

ASSIGNMENT

In this assignment you have to calculate the binding energy per nucleon for three stable nuclides of different masses. By calculating the values of each term in the equation, you will be able to see the relative importance of the various contributions to $\Delta E / A$.

4.18 Complete the table below; use the values for coefficients a, b, c, and d from Table 4.1.

Table 4.2   Contributions to nuclear binding energy

Nuclide	Z	N	A	Volume term/MeV	Surface term/MeV	Coulomb term/MeV	Symmetry term/MeV	ΔE MeV	$\Delta E/A$ MeV/nucleon
$^{7}_{3}$Li									
$^{56}_{26}$Fe									
$^{209}_{83}$Bi									

If you can program in BASIC, then write a program to calculate values of the quantities in Table 4.2, and use these to plot graphs similar to Figs 4.5 and 4.6. The *Nuclides database* can also be used to plot a binding energy graph (see Appendix A).

4.6 SUMMING UP

In this chapter you have looked at the information we can get from our knowledge of mass defects and binding energies. In the next chapter you can find out more about how this information enables us to make predictions about the stability of different nuclides and to calculate the energy released when an unstable nucleus disintegrates.

The knowledge we have developed so far is shown in summary form in Fig 4.7. This is a 'map' of binding energy per nucleon, shown on an N–Z plot. The figure shows contours of binding energy per nucleon, $\Delta E / A$. The most tightly bound nuclides have the greatest values of $\Delta E / A$. These, as shown in Fig 4.5, are in the region of $Z \sim N \sim 30$.

Fig 4.7 also incorporates the information from Fig 3.4, about the distribution of unstable nuclides about the line of stability.

If you think of the lines on the figure as being like depth contour lines on a map, and the most heavily shaded region as being the lowest part of the map, perhaps you will understand why this diagram is sometimes referred to as the 'valley of stability'.

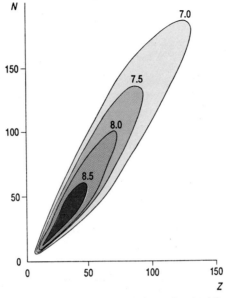

Fig 4.7 A contour map of the nuclear 'valley of stability.'

QUESTIONS

Table 4.3

Particle or nuclide	Mass/u
$_1^1 \text{p}$	1.007 276
$_0^1 \text{n}$	1.008 665
$_6^{15}\text{C}$	15.007 306
$_7^{15}\text{N}$	14.996 265
$_8^{15}\text{O}$	14.998 680
$_8^{16}\text{O}$	15.990 523
$_8^{17}\text{O}$	16.994 741
$_8^{18}\text{O}$	17.994 768

These questions require you to use the knowledge you have gained in this chapter to calculate mass defects and binding energies, and to use the results of such calculations to make predictions about the behaviour of certain nuclides. You will need to use the data in Table 4.3. All masses are nuclear (not atomic) masses.

4.19 Three nuclides having $A = 15$ are $_6^{15}\text{C}$, $_7^{15}\text{N}$ and $_8^{15}\text{O}$. Two of these are unstable. Calculate the mass defect (in u) and binding energy (in MeV) for all three. Which nuclide is most tightly bound? What decay processes might you expect to observe for the two nuclides which are unstable?

4.20 Calculate the binding energy per nucleon for each of the three stable isotopes of oxygen ($A = 16, 17, 18$) shown in Table 4.3. Use your answer to predict which isotope is the most abundant. Check your answer in a table of isotopes.

SUMMARY

The mass of a nucleus is less than the sum of the masses of the nucleons of which it is made. The difference is called the mass defect, ΔM. This is because the nucleus is in a lower energy state than are separate nucleons. Energy is required to separate a nucleus into its constituent nucleons. The energy required is called the binding energy, ΔE. The two are related by $\Delta E = \Delta M c^2$.

The binding energy per nucleon is greatest for nuclei with $A \sim 60$. Its dependence on A may be accounted for by the semi-empirical binding energy equation, whose terms relate to the liquid drop model of the nucleus.

Chapter 5

RADIOACTIVE DECAY

In this chapter you will use the knowledge you gained in the last chapter to understand one of the rules which govern radioactive decay. In the course of this discussion you will find out about the energy released when an unstable nucleus decays, and meet two unusual particles – the neutrino and the antineutrino.

LEARNING OBJECTIVES

After studying this chapter you should be able to:

1. write equations for nuclear decays involving α, β^+ and β^- emission, and electron capture;

2. calculate values of the energy Q released when an unstable nucleus decays;

3. determine, from values of Q, whether or not a particular radioactive decay is possible.

5.1 A CONDITION FOR DECAY

We will begin this section by considering a particular example of radioactive decay, then see how we can use our knowledge of the relationship between mass and energy to explain why this decay occurs.

Consider nuclide, $^{215}_{85}\text{At}$, which decays by α emission:

$$^{215}_{85}\text{At} \rightarrow {}^{211}_{83}\text{Bi} + {}^{4}_{2}\text{He}$$

This equation can be rewritten with masses in place of symbols (in this case, the values given are atomic masses rather than nuclear masses; this is not a problem, since there is the same number of electrons on each side of the equation):

$$214.998\ 66 \rightarrow 210.987\ 29 + 4.002\ 60 = 214.989\ 89$$

You can see that there is less mass after the decay than before. The decrease in mass is $(214.998\ 66 - 214.989\ 89) = 0.008\ 77$ u.

QUESTIONS

5.1 If this 'lost mass' appears as energy, how much energy is available from each parent nucleus which decays? (Give your answer in MeV.)

5.2 Use a table of nuclides to find the kinetic energy of the α particle emitted in this decay. Why is this value slightly less than the answer to the previous question? Where do you think the rest of the energy goes?

A decay which cannot occur

Can the nucleus $^{28}_{14}\text{Si}$ decay by α emission? The equation for this decay, and the mass equation, would be:

(a) $Q > 0$; decay possible

(b) $Q < 0$; decay impossible

Fig 5.1 **(a)** Radioactive decay (e.g. α emission) can occur if the mass of the daughter particles is less than the mass of the parent nuclide. **(b)** If mass increases in the proposed decay, Q is negative, and decay cannot occur.

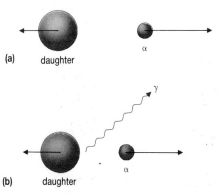

(a)   daughter

(b)   daughter

Fig 5.2 α emission, **(a)** without and **(b)** with, emission of an accompanying γ photon.

$$^{28}_{14}\text{Si} \quad \rightarrow \quad ^{24}_{12}\text{Mg} \quad + \quad ^{4}_{2}\text{He}$$

$$27.976\,93 \rightarrow 23.985\,05 + 4.002\,60 = 27.987\,65$$

Here you can see that there is more mass after the proposed decay than before it. This decay cannot therefore occur (unless some external source of energy is made available).

A general condition

A nuclear decay can occur if energy is released in the process, as in the first example above. We use the symbol Q to represent the energy released. We can then write:

parent nuclide → daughter nuclides + Q,

and the relationship between the masses of the nuclides and Q is:

Q = (parent mass – daughter masses) × c^2.

A condition for the decay to occur is simply:

$Q > 0$.

This is just one of the necessary conditions for nuclear decay to be possible. It is always true that, if Q is less than zero, the decay cannot occur. Fig 5.1 shows the 'mass–energy balance' picture of this.

We have restricted our discussion to α emission; however, the same rule also applies to β⁻ emission (see Section 5.2).

Where does the energy go?

The general equation for α decay is:

$$^{A}_{Z}\text{X} \rightarrow ^{A-4}_{Z-2}\text{Y} + ^{4}_{2}\text{He} + Q$$

The α particle is emitted with kinetic energy $E_{k\alpha}$. The daughter nucleus Y also has kinetic energy (E_{kY}), since it recoils when the α particle is emitted. In some cases the energy Q appears entirely in the form of the kinetic energy of these two particles. In other cases a γ photon of energy E_γ may also be emitted. These two possibilities are illustrated in Fig 5.2.

We can write:

$$Q = E_{k\alpha} + E_{kY} + E_\gamma$$

In the assignment which follows you are asked to investigate the way in which the energy Q is shared between the two particles.

ASSIGNMENT

During α emission by a parent nucleus, energy Q is released. Consider the case shown in Fig 5.2(a), in which no γ photon is emitted. We will take an imaginary case in which a parent nucleus X of mass 244 u emits an α particle of mass 4 u, leaving a daughter nucleus Y of mass 240 u , sixty times the mass of the α particle. (These numbers are integers, for convenience – although this implies no energy is released.)

Suppose we measure the velocity of the α particle and find it to be 1.8×10^7 m s⁻¹.

5.3 Use considerations of momentum to show that the recoil velocity of Y is 3×10^5 m s⁻¹, one sixtieth of the velocity of the α particle.

5.4 Calculate the kinetic energies of both Y and α, in MeV. (Take the mass of one nucleon as 1.67×10^{-27} kg.)

5.5 Show that the α particle has sixty times more energy than Y.

From these results you can see that the emitted α particle moves much faster than the daughter nucleus Y, and carries away most of the kinetic energy released in the decay process. It is rather like firing a cork from a pop-gun. The cork flies out fast, and the gun recoils slightly.

5.2 DECAY EQUATIONS

particle	:	alpha
other name	:	helium nucleus
symbols	:	α, $^{4}_{2}\text{He}_{2}$
charge	:	$+2e$, $+3.204 \times 10^{-19}$ C
rest mass	:	m_{α}; 6.647×10^{-27} kg
		4.002 604 u
		3726.4 MeV
lifetime	:	stable

particle	:	electron
other names	:	beta particle
symbols	:	e^{-}, $^{0}_{-1}e_{0}$, β^{-}
charge	:	$-e$, -1.602×10^{-19} C
rest mass	:	m_{e}; 9.109×10^{-31} kg
		0.000 549 u
		0.511 MeV
lifetime	:	stable

particle	:	neutrino
symbol	:	$\overline{\nu}$
charge	:	0
rest mass	:	0
lifetime	:	stable

antiparticle	:	antineutrino
symbol	:	$\overline{\nu}$
charge	:	0
rest mass	:	0
lifetime	:	stable

In this section we will look in more detail at the equations for nuclear decay and think about the energy released. We will look at α and β decays and consider three less familiar nuclear processes: neutron emission, proton emission and electron capture.

We will write general equations for these processes. The parent nucleus is X the daughter nucleus is Y.

α emission

Proton-rich nuclei may become more stable by emitting an α particle. The general equation for this (Section 5.1) is:

$$^{A}_{Z}\text{X}_{N} \rightarrow ^{A-4}_{Z-2}\text{Y}_{N-2} + ^{4}_{2}\text{He}_{2} + Q \qquad (1)$$

You should check that there is a balance of Z (and hence charge) and A on the two sides of this equation. (The α particle is written as $^{4}_{2}\text{He}_{2}$ to make the accounting easier.) Z decreases by 2 and A decreases by 4. The equations which follow below must also balance in this way, since charge and mass number must both be conserved. The energy released, Q, is included on the right-hand side to ensure that [mass + energy] is conserved.

β⁻ emission

Neutron-rich nuclei tend to decay by β⁻ emission. A β⁻ particle is simply an electron. The general equation is:

$$^{A}_{Z}\text{X}_{N} \rightarrow ^{A}_{Z+1}\text{Y}_{N-1} + ^{0}_{-1}e_{0} + \overline{\nu} + Q \qquad (2)$$

This equation has an extra particle in it, the **antineutrino**, $\overline{\nu}$. This is a particle with no mass and no charge, so it does not affect the balance of the quantities in the equation. We will come to the reason for including this particle shortly.

Check the balance of Z and A. The number of neutrons has decreased by one, the number of protons has increased by one. It is as if a neutron had decayed to become a proton and an electron:

$$^{1}_{0}\text{n}_{1} \rightarrow ^{1}_{1}\text{p}_{0} + ^{0}_{-1}e_{0} + \overline{\nu} + Q \qquad (3)$$

ASSIGNMENT

Equation (3) suggests that a neutron is unstable and will decay by β⁻ emission. Can this be true?

5.6 Use values for the masses of the particles to determine whether there is a decrease in mass in the course of such a neutron decay. Find the value of Q in MeV.

5.7 Use a table or chart of nuclides to find out the half-life of a free neutron. (This value does not apply to neutrons within nuclei.)

particle	:	proton
symbols	:	p, $_1^1$p$_0$
charge	:	$+e$, $+ 1.602 \times 10^{-19}$ C
rest mass	:	m_p; 1.673×10^{-27} kg
		1.007 276 u
		938.3 MeV
lifetime	:	stable

Why the antineutrino? In the next chapter you will see some of the evidence for the existence of the antineutrino. For the moment, its appearance in equations 5.2 and 5.3 can be explained as follows.

We have seen that charge must balance on both sides of these equations, as must the number of massive particles (protons and neutrons). In addition, the number of particles on the left must equal the number of particles on the right. If we wrote the equations without $\bar{v}$, there would be one more particle on the right than on the left. A particle would have been created during the decay. This cannot happen – the number of particles is conserved, in the same way that charge is conserved. We have to add an 'antiparticle', the antineutrino, on the right-hand side. Count +1 for every particle, and –1 for every antiparticle. In equation (3) there is one particle on the left, and [two particles + one antiparticle] on the right. The equation is balanced.

It should not surprise you to learn that there is a very similar particle, the neutrino, for which the symbol is v. This is a particle of matter; the antineutrino is **antimatter**. If we go on to consider the case of β^+ emission, you will become more familiar with the way in which these strange entities keep appearing in nuclear processes.

particle	:	neutron
symbols	:	n, $_0^1$n$_1$
charge	:	0
rest mass	:	m_n; 1.675×10^{-27} kg
		1.008 665 u
		939.6 MeV
lifetime	:	10.8 min (β^- decay)

β^+ emission

In Section 3.2 you saw that many proton-rich nuclides are β^+ emitters. Here is the general equation:

$$_Z^A X_N \rightarrow _{Z-1}^A Y_{N+1} + _1^0 e_0 + v + Q \tag{4}$$

The β^+ particle is shown as $_1^0 e_0$. It is an antiparticle, a piece of antimatter. Its mass is the same as that of an electron and it has positive charge, $+e$.

A proton-rich nuclide can become more stable if it reduces the number of protons by one and increases the number of neutrons by one. To do this, of course, it must emit a positively charged particle. The underlying process thus seems to be:

$$_1^1 p_0 \rightarrow _0^1 n_1 + _1^0 e_0 + v + Q \tag{5}$$

You can now check that this equation is balanced in terms of three quantities: charge (Z), mass number (A) and the numbers of particles. Remember to count the β^+ as an antiparticle and the neutrino as a particle.

QUESTION

You can use equation (5) to investigate the stability of free protons. Use values of the masses of these particles to check the mass–energy balance of this decay process.

5.8 Is a free proton an unstable particle?

antiparticle	:	positron
other names	:	antielectron
symbols	:	e$^+$, $_{+1}^0$e$_0$, β^+
charge	:	$+e$, $+1.602 \times 10^{-19}$ C
rest mass	:	m_e; 9.109×10^{-31} kg
		0.000 549 u
		0.511 MeV
lifetime	:	stable

Electron capture

This is a third kind of radioactive decay, in which an unstable nucleus becomes more stable by absorbing one of the electrons which orbit it. Since the electrons closest to the nucleus constitute the K shell, and it is usually one of these which is absorbed, this process is sometimes called K capture. The general equation is:

$$_Z^A X_N + _{-1}^0 e_0 \rightarrow _{Z-1}^A Y_{N+1} + v + Q \tag{6}$$

As usual, you can check that this equation is appropriately balanced. In the questions which follow you are asked to think about the process underlying electron capture, in the same way that we have discussed β^-, β^+ and α decay.

5.9 Would you expect electron capture to be exhibited by proton-rich or neutron-rich nuclei?

5.10 Write an equation similar to equations (3) and (5) to show the underlying process.

5.11 Explain how this equation is balanced in terms of charge, mass and numbers of particles.

5.12 Earlier, in questions 5.6 and 5.7, you showed that a free neutron is unstable and can decay into a proton and an electron. Consider your equation for electron capture. Why might you expect this process to be unlikely to occur?

Other decay processes

We have looked at decays involving α particles and electrons. There are some other types of radioactive decay observed in practice. These generally involve nuclides produced artificially in nuclear reactors. The processes involved are natural, but the nuclides concerned are not found in nature on our planet.

Proton and neutron emission. In a nuclear reactor, amongst the many nuclides formed may be some which are very proton-rich or very neutron rich. To become more stable, it is possible for these to emit a single nuclide, either a proton or a neutron. You can write your own general equations for these processes. Here is an example of proton emission:

$$^{5}_{3}\text{Li}_2 \rightarrow {}^{4}_{2}\text{He}_2 + {}^{1}_{1}\text{p}_0 + Q \tag{7}$$

It should not surprise you, if you look back to the binding energy graph of Fig 4.5, that this isotope of lithium becomes more stable when it decays to $^{4}_{2}\text{He}_2$.

If you have a sufficiently detailed chart of nuclides you will be able to find more examples. You should be able to guess where to find proton emitters and neutron emitters, relative to the line of stability.

Nuclear fission. Some massive nuclei become more stable by splitting into two quite large fragments, rather than by emitting a small particle such as a proton, neutron, electron or alpha particle. This is termed nuclear fission, and is the basis of nuclear power generation and many nuclear weapons. You can understand nuclear fission on the same basis as radioactive decay; this is dealt with in Chapter 8.

Nuclear fusion. Some light nuclei can combine or fuse to form larger, more stable, nuclei. Energy is released in the process. The conditions under which this can occur and the uses made of this process are discussed in Chapter 9.

ASSIGNMENT

Complete the table below:

Decay:	Effect on parent nuclide:		
	Z	N	A
α emission	decreases by 2	decreases by 2	decreases by 4
β^- emission			
β^+ emission			
e capture			
p emission			
n emission			

In the next chapter we will look at the underlying mechanisms of these processes; the next section shows you how you can use the ideas of mass and energy developed in Chapter 4 to calculate the energy released in radioactive decay processes.

5.3 CALCULATING Q

We have looked at the equations for radioactive decay and established the condition for decay to be possible, namely that Q must be greater than zero. Now we can calculate Q from our knowledge of nuclear masses. There is one point to beware of – the mass values quoted in tables are the masses of atoms. You must take account of what is happening to the atomic electrons during radioactive processes.

Counting the electrons

The problem is that, during radioactive decay, the charge of the nucleus changes. To retain charge neutrality (so that we can use atomic masses), we must ensure that the number of electrons changes to match the change in the nuclear charge. There is usually no problem, but with β^+ emission we must take care.

α emission
Think about equation (1):

$$_Z^A X_N \rightarrow _{Z-2}^{A-4} Y_{N-2} + _2^4 He_2 + Q$$

If we consider X as an atom (rather than as a nuclide), it must have Z electrons. The daughter atom Y has Z–2 electrons. We must now think of the α particle as a helium atom, which has two electrons.

Clearly the numbers of electrons on the two sides of the equation are the same. We can safely use atomic masses to calculate Q without any problem:

$$Q = (M_X - M_Y - M_\alpha)\,c^2 \qquad (8)$$

β⁻ emission
We have seen that the equation for this is:

$$_Z^A X_N \rightarrow _{Z+1}^A Y_{N-1} + _{-1}^0 e_0 + \bar{\nu} + Q$$

The parent atom has Z electrons; the daughter atom Y has Z+1 electrons. The nucleus has become more positively charged. We need an extra electron from somewhere to maintain neutrality. Fortunately there is an electron on the right-hand side – the β^- particle. (Remember that the anti-neutrino has no mass.) So the equation is balanced, and we can write (using atomic masses):

$$Q = (M_X - M_Y)\,c^2 \qquad (9)$$

β⁺ emission
This is where problems might arise. Consider equation (4):

$$_Z^A X_N \rightarrow _{Z-1}^A Y_{N+1} + _1^0 e_0 + \nu + Q$$

There are Z electrons orbiting the nucleus of the X atom; there are only Z–1 electrons in the Y atom. This means that an extra electron is released on the right-hand side, as well as the positron emitted by the nucleus. Each of these has mass m_e. To take account of this, we must write:

$$Q = (M_X - M_Y - 2m_e)\,c^2 \qquad (10)$$

Fig 5.3 on the next page represents this pictorially.

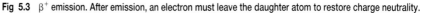

Fig 5.3 β⁺ emission. After emission, an electron must leave the daughter atom to restore charge neutrality.

Electron capture

There is no problem here. The nucleus captures one of the electrons which are orbiting it. The nuclear charge decreases by one, and the number of orbiting electrons also decreases by one. Q is given by:

$$Q = (M_X - M_Y)\,c^2$$

Two sample calculations

In question 5.1, you calculated the energy released in a particular case of α decay. Here are two further examples of such calculations of energy released.

1 $^{90}_{38}$Sr is a well known β⁻ emitter; your laboratory sources may include a sample of this isotope of strontium. It decays to $^{90}_{39}$Y. We can write the equation for this:

$$^{90}_{38}\text{Sr} \rightarrow {}^{90}_{39}\text{Y} + {}^{0}_{-1}e_0 + \bar{\nu} + Q$$

The energy released for each nucleus which decays is:

$$\begin{aligned} Q &= (M_{Sr} - M_Y)\,c^2 \\ &= (89.9073 - 89.9067) \times 931.3 \\ &= 0.559 \text{ MeV} \end{aligned}$$

2 Here is the equation for the decay of an isotope of zinc:

$$^{65}_{30}\text{Zn} \rightarrow {}^{65}_{29}\text{Cu} + {}^{0}_{+1}e_0 + \nu + Q$$

Using equation (10) we can calculate Q.

$$\begin{aligned} Q &= (M_{Zn} - M_{Cu} - 2m_e)\,c^2 \\ &= (64.9292 - 64.9278 - [2 \times 0.000\,55]) \times 931.3 \\ &= 0.279 \text{ MeV} \end{aligned}$$

QUESTIONS

nuclide	mass/u
$^{7}_{4}$Be	7.016 93
$^{7}_{3}$Li	7.016 01
$^{64}_{28}$Ni	63.927 96
$^{64}_{29}$Cu	63.929 76
$^{64}_{30}$Zn	63.929 15
$^{224}_{88}$Ra	224.020 22
$^{228}_{90}$Th	228.028 75

In answering the questions which follow, you will need to use some or all of the following values of atomic masses:

5.13 The nuclide $^{228}_{90}$Th is an α emitter in the 4n series shown in Fig 3.6 (page 39). Calculate the energy Q released when one such nucleus decays.

5.14 The nuclide $^{64}_{29}$Cu is unusual; it can decay by β⁻ or β⁺ emission, or by electron capture. Calculate the value of Q in each case.

5.15 Can the nuclide $^{7}_{4}$Be decay by β⁺ emission? (Hint: determine Q for this decay. Is the answer positive or negative?)

5.4 PICTURING RADIOACTIVE DECAY

We have established a condition for radioactive decay to occur: there must be a release of energy Q during the decay. This happens if the total mass of the particles after the event is less than the mass before the event. (Remember: lower mass means lower energy.)

Radioactive decay is the way in which unstable nuclides can reach a more stable, lower energy state. Energy is released in the process. The binding energy per nucleon increases; in other words, the nucleons are more tightly bound together after the nucleus has decayed.

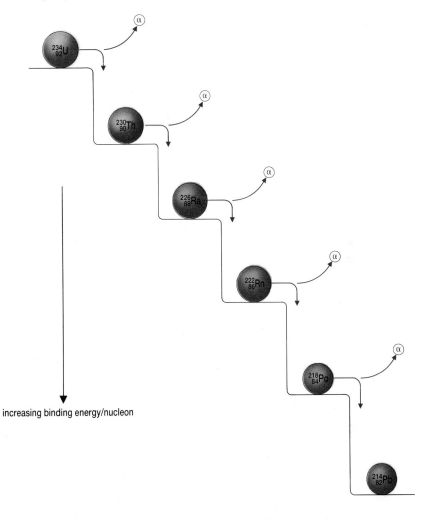

increasing binding energy/nucleon

Fig 5.4 The increase in binding energy per nucleon down the 4n+3 decay series, represented as a series of steps.

ASSIGNMENT

A decay series is a sequence by which a large, unstable nucleus can achieve stability. Table 5.1 shows six nuclides of the 4n+2 decay series, together with the values of their binding energies. Each nuclide decays to the next by α emission.

Table 5.1   Binding energies in the 4n+2 decay series

Nuclide	ΔE/MeV	$\Delta E \div A$/MeV nucleon^{-1}
$^{234}_{92}$U	1778.30	
$^{230}_{90}$Th	1754.86	
$^{226}_{88}$Ra	1731.34	
$^{222}_{86}$Rn	1707.94	
$^{218}_{84}$Po	1685.24	
$^{214}_{82}$Pb	1663.00	

Complete the table by calculating the binding energy per nucleon for each nuclide.

5.16 How does the binding energy per nucleon change down the decay series?

Representing a decay series

There are several ways in which we can picture the increase in binding energy per nucleon during radioactive decay. Here are three figures which show three different representations of the same information.

Fig 5.4 shows the way in which binding energy per nucleon changes in such a decay series.

Fig 5.5 The decay series gradually approaches the line of stability.

Fig 5.5 shows part of the binding energy curve (refer back to Fig 4.5) with the 4n+2 series marked on it to show how the binding energy per nucleon increases down the series. The nuclides gradually approach the line of stability.

Fig 5.6 shows the same decay series, this time superimposed on the 'valley of stability' of Fig 4.7. The series appears to slide 'downhill' into the region of greatest stability.

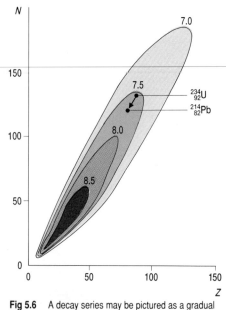

Fig 5.6 A decay series may be pictured as a gradual slide into the 'valley of stability'.

Unanswered questions

The diagrams which you have just been looking at should help to summarise the ideas we have developed in this chapter, but you may have some questions about radioactive decay which remain unanswered.

We have established that decay can occur if Q is positive; that is, if energy is released in the process. But does decay always occur in practice if Q is positive?

Many nuclides can decay by α emission and become more stable in the process. And yet, this is not an instantaneous effect; α emitters can have half-lives shorter than 10^{-6} s, or longer than 10^{15} years. This is a variation by a factor of 10^{29}. Why the difference? Why don't these nuclides decay instantaneously, if it would make them more stable?

Why does a decay series, such as the 4n+2 series shown in Figs 5.4, 5.5 and 5.6, stop at one particular nuclide? Why does it not continue to decay to yet more stable nuclides, farther down into the valley of stability?

Try to think about these questions now. They may help to clarify your ideas about the concepts we have looked at in this chapter. If you can at least understand these questions (without yet being able to answer them), then you should be able to understand the next chapter. There we will look at some explanations of the mechanisms of nuclear decay processes, which should help us to provide answers to these questions.

SUMMARY

For radioactive decay to occur it is necessary for the total mass of the daughter particles to be less than the mass of the parent nucleus. In radioactive decay, A, Z and N are conserved; the number of particles is also conserved. It is necessary to postulate the existence of neutrinos and antineutrinos to ensure that particle numbers are conserved in β decay.

The energy released in radioactive decay may be deduced (by using $E = mc^2$) from the decrease in mass. It appears as kinetic energy of the particles, and sometimes also as a γ photon.

Chapter 6

THE PROCESSES OF RADIOACTIVE DECAY

So far in this theme we have looked at radioactive decay as a phenomenon. We know that some nuclides are unstable, and that they decay to become more stable. We have looked at the energy and mass changes involved in alpha and beta emission. Now we must look at the underlying mechanisms involved in these processes.

LEARNING OBJECTIVES

After studying this chapter you should be able to:

1. describe appropriate models to explain the mechanisms of α, β^+, β^- and γ emission, and electron capture;

2. describe the evidence which supports these models;

3. outline the evidence for the existence of neutrinos.

ASSIGNMENT

In this chapter you will need to make use of many of the ideas about forces and energy in the nucleus developed in Chapter 2. You should recall the following concepts:

> nuclear strong force
> Coulomb repulsion between protons
> nuclear potential well
> Coulomb barrier
> liquid drop model

If you are not clear about the significance of any of these concepts then scan through Chapter 2, and your own notes, to remind yourself of their meanings.

6.1 ALPHA EMISSION

So far we know the following:

1. An α particle consists of two protons and two neutrons; it is a helium nucleus, $^4_2$He.

2. This is an especially stable configuration of nucleons.

3. α particles are emitted by high-mass, proton-rich nuclei.

4. From the Rutherford–Geiger–Marsden experiment we know that fast-moving α particles do not appear to penetrate the nuclei of gold (and other) atoms.

The question we are now going to address is this: we know that, when a nucleus emits an α particle, it becomes more stable. It achieves a lower energy state. So why is the process not instantaneous? To begin to answer this we must look at the observed energies of α particles emitted by different nuclides.

Alpha particle energies

The velocity, and hence the kinetic energy, of an α particle may be found from its deflection in a magnetic field. Another indication of the energies of α particles can be found from their tracks in a cloud chamber. Fig 6.1 shows such tracks. These are long, straight lines which end rather abruptly. How are they formed?

(a)   (b)

Fig 6.1   Tracks of a particles in a cloud chamber.

As an α particle travels through the air in the cloud chamber it ionises air molecules with which it collides along its path. Droplets form around the ions and the α track is revealed.

The α particle gradually loses energy in the process; about 30 eV for each collision. When it has lost all of its energy, it comes to a halt. (It will become a helium atom.) Longer tracks indicate greater energy.

QUESTIONS

6.1 Fig 6.1(a) shows the tracks of α particles from one nuclide. They are all of the same length. What does this tell you about the range of energies of the α particles?

6.2 Fig 6.1(b) shows the tracks of α particles from a different nuclear species. What can you say about the energies of these α particles?

ASSIGNMENT

In this assignment you are asked to look at the relationship between α particle energies and radioactive half-life, $t_{1/2}$. Table 6.1 lists these two quantities for the α emitting nuclides of the 4n decay series.

Table 6.1   α emitters of the 4n decay series

Nuclide	$E_{k\alpha}/\text{MeV}$	$t_{1/2}/\text{s}$
$^{232}_{90}\text{Th}$	4.0	4.5×10^{17}
$^{228}_{90}\text{Th}$	5.4	6.0×10^{7}
$^{224}_{88}\text{Ra}$	5.7	3.1×10^{5}
$^{220}_{86}\text{Rn}$	6.3	52
$^{216}_{84}\text{Po}$	6.8	0.16
$^{212}_{84}\text{Po}$	8.8	3.0×10^{-7}

6.3 If you inspect this table you may be able to see a relationship between $E_{k\alpha}$ and $t_{1/2}$. What qualitative relationship can you see between these two quantities?

Getting out of the nucleus

How can we describe this relationship between the energy of emitted α particles and the half-life of the nuclide concerned? Nuclides which decay quickly (short half-life) produce energetic α particles. Or, to put it the other way around, energetic α particles find it easier to get out of the nucleus.

This suggests that the α particle is formed inside the nucleus and can escape from the nucleus, but only with some difficulty. For a nuclide such as $^{232}_{90}$Th (in Table 6.1), for which $t_{1/2} = 1.4 \times 10^{10}$ years, it seems that an α particle can only escape with very great difficulty.

To develop our understanding of this, we must go back to some of the ideas of Chapter 2. There we described the nucleus as being somewhat like a crater. The particles of which it is made are trapped inside the crater, and particles fired from the outside (such as Geiger and Marsden's α particles) find it hard to get in.

Now we can picture the α particle which is emitted in radioactive decay as being trapped inside this crater. It has several MeV of energy – we know this because of the speed with which it moves when it escapes – but it has not got enough energy to climb out of the crater.

Since the tracks of α particles show that they all have the same energy, we deduce that they occupy a particular energy level within the nucleus. This is illustrated in Fig 6.2 (i.e. Fig 2.7(c) with the energy level of the α particle added). The α particle cannot escape past the barrier; it is held in the nucleus by the strong nuclear attraction of the other nucleons.

Perhaps you think we have just replaced one problem with another. How can we picture the process by which an α particle does eventually escape from the nucleus?

You might guess that the α particle somehow gets enough energy to escape over the top of the barrier. But this cannot be the case; if it were, the α particles would be found to have energies corresponding to the top of the barrier. In fact, it turns out that they manage to pass right through the barrier. We imagine that the α particle exists within the nucleus; it is energetic and rushes about, frequently bouncing off the walls of the crater. There is a slim chance that it will escape through the walls; this is an effect of quantum mechanics, the mechanical laws obeyed by nucleons. (We do not observe such behaviour in our macroscopic world. If you rush about the room, bouncing off the walls at frequent intervals, you would not expect to appear suddenly in the room next door.) The passing of a particle through an apparently impenetrable barrier is known as **quantum mechanical tunnelling**.

Fig 6.2 The energy level of an α particle within the nuclear potential energy well.

Energy and half-life

Fig 6.3 shows how we can picture an explanation for the observed above relationship between $E_{k\alpha}$ and $t_{1/2}$, which you looked at in the last assignment. The figure shows two different α particle energy levels within a nucleus. The more energetic particle has to penetrate a thinner region of the

Fig 6.3 α particles have to penetrate the Coulomb barrier to escape from the nucleus.

barrier than the lower, less energetic α particle. It thus escapes more readily, and so the corresponding half-life is shorter.

(a)   (b)   (c)

Fig 6.4 The escape of an α particle from the nucleus
(a) the α particle appears at the nuclear surface
(b) it pulls clear of the nucleus
(c) it becomes completely free.

Another picture

The description of α emission given above is in terms of energies. We can think of the process in another way, in terms of the forces involved. Fig 6.4 shows what happens.

The α particle is a stable unit; we picture it existing and moving about within the nucleus. An energetic α particle may try to escape when it reaches the surface of the nucleus, but it is pulled back by the strong attraction of the other nucleons, both protons and neutrons. There is only a small chance that it will break away from the nucleus. Once outside, it is repelled away by the Coulomb repulsion between protons.

6.2 BETA DECAY

Fig 6.5 The three particles produced in β⁻ decay.

There are two kinds of β emission. For β⁻ emission we know the following:

1. β⁻ is exhibited by neutron-rich nuclei.

2. A β⁻ particle is an electron.

3. β⁻ emission is accompanied by the emission of an antineutrino, $\bar{\nu}$.

The process is illustrated in Fig 6.5. You should be able to write equivalent statements for β⁺ emission.

THE PROCESSES OF RADIOACTIVE DECAY   **65**

Table 6.2	Nuclides with $A = 101$	
Nuclide	Z	M/u
$^{101}_{40}$Zr	40	100.9200
$^{101}_{41}$Nb	41	100.9138
$^{101}_{42}$Mo	42	100.9089
$^{101}_{43}$Tc	43	100.9059
$^{101}_{44}$Ru	44	100.9041
$^{101}_{45}$Rh	45	100.9047
$^{101}_{46}$Pd	46	100.9068
$^{101}_{47}$Ag	47	100.9913
$^{101}_{48}$Cd	48	100.9173

6.9 Which nuclide has the lowest mass, and is therefore the most stable? Indicate this on your graph.

6.10 Which nuclides are neutron-rich, and will therefore decay by α emission towards the minimum mass? Indicate these transitions on your graph by arrows labelled β^-.

6.11 Indicate β^+ transitions by proton-rich nuclides towards the minimum. You can explore other such sets of nuclides (having other values of A) using the *Nuclides database* (see Appendix A).

Energies of β particles

It is possible to measure the velocities of β particles, and hence their kinetic energies, in a similar way to those of α particles, by observing their deflection in a magnetic field. The results for β particles are very different to those for α particles.

Fig 6.6 Spectra of β particles **(a)** a typical spectrum **(b)** the spectra for $^{64}_{29}$Cu, an emitter of both β^- and β^+ particles.

For α particles we saw that all particles emitted by one nuclide have the same value of kinetic energy (or two, or a few, values). This is not the case for β particles. They have a range, or spectrum, of energies. This is shown in Fig 6.6(a).

This graph shows that the energy of β particles lies between zero and some maximum value. This maximum is the Q value for the decay. This tells us that only a few of the emitted β particles get all of the available energy from the nuclear decay. Where does the rest of it go?

Fig 6.6(b) shows the β spectra for $^{64}_{29}$Cu. This nuclide decays by both β^- and β^+ emission, so there are two spectra. In question 5.14 you calculated the Q values for these two decays. You should check to confirm that the cut-off energies of the two spectra agree with the Q values which you calculated.

Fig 6.7 summarises the way in which α and β particles are deflected in a magnetic field.

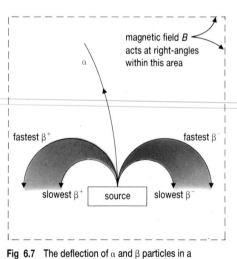

Fig 6.7 The deflection of α and β particles in a magnetic field.

Evidence for neutrinos

The emitted β particle does not account for all of the energy released in β decay. Nor does the kinetic energy of the daughter nucleus explain the discrepancy. There must be some other particle which carries off the rest of Q.

THE PROCESSES OF RADIOACTIVE DECAY

(a)

(b)

Fig 6.8 The paths of the products of radioactive decay (a) α decay (b) β decay.

(a)

(b)

photon

(c)

Fig 6.9 An atom, **(a)** before, **(b)** during, and **(c)** after electron capture by the nucleus.

This third particle is the antineutrino (in the case of β⁻ decay) which we have discussed before. Its existence was originally suggested to explain the shape of the β⁻ spectrum, Fig 6.6(a). Other pieces of evidence helped to confirm its existence.

Experiments were performed in the early 1920s to measure the energy released in radioactive decay. A source was enclosed in a thick-walled calorimeter from which β⁻ particles could not escape. The rise in temperature allowed the energy released during each decay to be determined. This was found to be less than Q; rather, it was equal to the average energy of the β⁻ particles. Something was escaping from the calorimeter and taking energy with it.

Later experiments with β⁻ sources in cloud chambers showed that the daughter nucleus did not recoil in precisely the opposite direction to the β⁻ particle (see Fig 6.8). Again, another unseen particle was needed to explain this observation.

Neutrinos are very difficult to observe directly, because they interact only very weakly with matter. They have been described as follows. Picture yourself in an aircraft, flying high over a rugby ground. You can see the players moving about, but you are too high to see the ball. Provided that you know there is a ball involved in the game of rugby, you can probably work out its movements from observations of the players' movements. If you know the rules of the game, you can probably make even better guesses about the ball. The neutrino is like the rugby ball; for a long time its existence was inferred from observations of the behaviour of other particles.

At the end of this chapter you can read about some of the direct evidence which has been collected to show that neutrinos and antineutrinos really do exist.

The weak interaction

The force responsible for β decay is the nuclear weak interaction, another of the four fundamental forces of nature. This interaction is important in the synthesis of the chemical elements in the stars.

Because this interaction is weak, the half-lives of β emitters are long (generally more than 1s, which is long in nuclear terms), and may be as much as 10^{14} years.

It is possible to relate β energies to half-lives, in the same way that we looked at these quantities for α decay. Again, we would generally find that high energy β particles are emitted by nuclides with short half-lives, though it is not possible to obtain such a simple graph.

Electron capture

In this process an electron from the K shell of the parent atom is absorbed by a proton-rich nucleus. Since a particle (the electron) disappears, a neutrino must be produced in the process. The questions which follow ask you to think about electron capture.

QUESTIONS

6.12 The daughter nuclide is very, very massive compared to the neutrino. Which of these two particles will have the greater share of the released energy Q? Give a reason for your answer.

6.13 A vacancy is left in the K shell of the daughter atom (see Fig 6.9). An electron from a higher level fills the vacancy, and a photon of radiation is emitted. What kind of radiation is this? Will it be characteristic of the parent atom or the daughter atom?

THE PROCESSES OF RADIOACTIVE DECAY

67

6.3 GAMMA EMISSION

So far in our exploration of the mechanisms of radioactive decay we have talked of the energy Q released being carried away as kinetic energy of the product particles. But, as you are no doubt aware, there is another possibility. Some of the energy may be released as γ radiation. γ rays make up the most energetic, short wavelength part of the electromagnetic spectrum; refer back to Fig 1.9 to remind yourself of where they fit into the spectrum.

To understand the process of γ emission, you need to be familiar with the way in which electromagnetic radiation such as visible light and X-rays results from changes of electron energies in atoms.

ASSIGNMENT

What follows is a very brief account of the production of photons by isolated atoms (see Fig 6.10).

Fig 6.10   A photon is emitted when an electron falls to a lower atomic energy level.

1.   The electrons of an isolated atom occupy narrowly defined energy levels.

2.   The number of electrons which can occupy any level is strictly limited.

3.   If there is a vacant site in an energy level, an electron from a higher level can 'fall' to occupy the vacancy.

4.   When it does this the electron's energy decreases, and a single photon is emitted.

The evidence for this description comes from the analysis of the spectrum of light emitted by hot gases. You have probably seen the spectra of sodium and mercury vapour lamps. These are line spectra.

Only certain narrowly defined frequencies of light are observed, and these may be related to the narrowly defined energy levels of the electrons of the atoms which make up the gas.

You should be familiar with the terms *line spectrum*, *energy level* and *photon*, and the equation $hf = E_{initial} - E_{final}$. If you are uncertain about these points, check them out now.

Gamma emission: the evidence

Many radioactive sources emit γ rays along with α or β particles. Tables of isotopes and some decay charts give information about the γ ray energies. The numbers quoted are generally the energies E_γ of γ photons, in MeV or keV. We will look at some of these values and try to understand what this information tells us about the nucleus.

The nuclide $^{226}_{88}$Ra decays to $^{222}_{86}$Rn by α emission. γ photons are detected, all of which have energy 0.19 MeV.

The nuclide $^{27}_{12}$Mg decays to $^{27}_{13}$Al by β emission. Three energies of γ photon are observed, having energies 0.18 MeV, 0.83 MeV and 1.01 MeV. (What do you notice about these three numbers?)

Other nuclides give several different γ photon energies. You can find examples in isotope tables and charts.

The important point to realise here is that all of these represent line spectra for the γ radiation. Only certain narrowly defined energies are present. By correspondence with the analysis of atomic spectra outlined above, this suggests that nuclei exist in narrowly defined energy states or levels. We will look at this in the next assignment. But first, here is a brief note about notation.

After α or β decay, the daughter nucleus Y may be left in an excited state, which we will denote Y*. This then decays to a stable lower state, the ground state, with the emission of a γ photon. For a parent nuclide X which decays by α emission, we write:

$$X \rightarrow Y^* + \alpha + Q'$$

$$\text{and } Y^* \rightarrow Y + \gamma$$

The energy Q' released in the α emission, and the energy E_γ of the γ photon, are related to the Q value of the whole decay process by:

$$Q = Q' + E_\gamma$$

ASSIGNMENT

In this assignment you can work out a way of representing radioactive decay including γ emission. One simple example is given, and then you can answer some questions on a more complicated example.

Fig 6.11(a) shows an energy level scheme for the α decay of $^{226}_{88}$Ra. The daughter nucleus may be formed in its ground state, or in an excited state. A γ photon of energy 0.19 MeV is emitted as the excited nucleus falls to the ground state.

6.14 Q for this decay is 4.78 MeV. Two energies of α particles are detected. The most energetic α particles have energy 4.78 MeV. What would you expect the energy of the other α particles to be? This example should show you why we sometimes observe cloud chamber tracks of two or more lengths, such as those we looked at earlier in Fig 6.1(b).

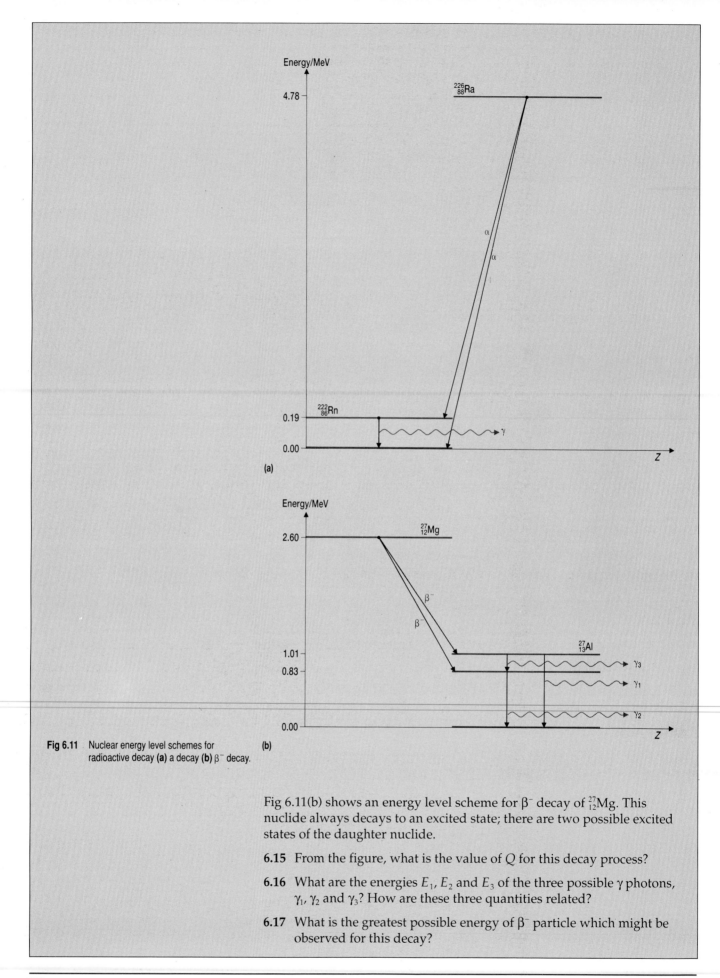

Fig 6.11 Nuclear energy level schemes for radioactive decay (a) a decay (b) β⁻ decay.

Fig 6.11(b) shows an energy level scheme for β⁻ decay of $^{27}_{12}$Mg. This nuclide always decays to an excited state; there are two possible excited states of the daughter nuclide.

6.15 From the figure, what is the value of Q for this decay process?

6.16 What are the energies E_1, E_2 and E_3 of the three possible γ photons, γ_1, γ_2 and γ_3? How are these three quantities related?

6.17 What is the greatest possible energy of β⁻ particle which might be observed for this decay?

THE PROCESSES OF RADIOACTIVE DECAY

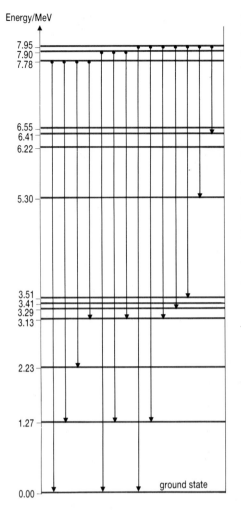

Energy/MeV

Fig 6.12 Nuclear energy levels and transitions for $^{31}_{15}$P.

Fig 6.12 shows the energy level diagram for the nuclide $^{35}_{10}$P, with some of the possible transitions indicated. This diagram has been deduced from the energies of γ photons. You can imagine the complicated γ ray spectrum which had to be interpreted in order to deduce this diagram.

Excited states and the shell model

How can we picture these excited states of a nucleus? There are several different models to consider; one of these is the shell model. This compares a nucleus to an atom. Electrons orbit the nucleus of an atom; they can be thought of as occupying energy levels. If the atom gains energy one or more of its electrons moves up to a higher energy level. (If you are studying chemistry, you will realise that this is a greatly simplified picture of what is now known about atoms.)

Similarly we can picture the nucleons of a nucleus as moving about within the nucleus, under the influence of the forces between them. An individual nucleon may increase its energy; we say it has moved up to a higher energy shell. The energies of the possible shells are shown as energy levels in Figs 6.11 and 6.12, which you studied earlier.

We can explain β decay using the shell model. Fig 6.13 shows the energy level diagram for an imaginary nuclide. The levels occupied by protons and neutrons are shown; no more than two protons and two neutrons may occupy any level.

This diagram represents a nuclide with two more neutrons than protons. That is, it is neutron-rich, since light nuclides prefer to have equal numbers of protons and neutrons. (Refer back to Fig 3.1 to remind yourself of this.) The nuclide can reduce its energy and become more stable by changing a neutron to a proton; this is the process of β⁻ decay. This is represented by the arrow on the diagram. Fig 6.13(b) shows the nucleon energies in the daughter nuclide. You can see that the overall energy of the nucleus is lower, and so the nucleus is more stable.

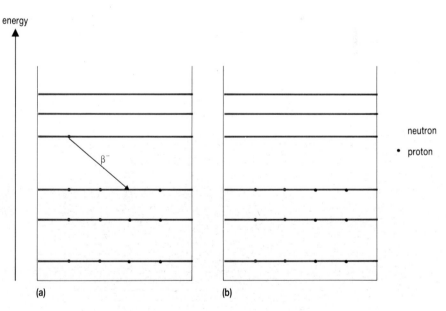

Fig 6.13 β⁻ decay for a neutron–rich nuclide.

There are other ways of describing the excited states of a nucleus. The shell model assumes that the nucleons retain their individual identity within the nucleus. Other models ignore the separate nucleons, and describe the behaviour of the nucleus as a whole. These are described as 'collective' models. Fig 6.14 shows some of the nuclear motions involved.

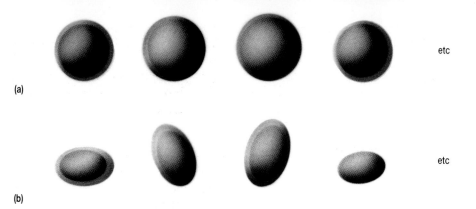

(a)

(b)

Fig 6.14 A large nucleus may **(a)** vibrate, or **(b)** rotate.

A large nucleus can change its shape. If it has enough energy it may vibrate rather like the wobblings of jelly on a plate. It can expand and contract – this is sometimes described as 'breathing'. Only certain modes of vibration are allowed, with specific values of energy.

Large nuclei are generally not spherical. A distorted sphere can rotate in different ways, and with different energies.

QUESTION

6.18 Fig 6.15 shows the energy level diagram for the β^- decay of $^{60}_{27}Co$.
Use the diagram to determine:
(a) the energies of emitted γ photons;
(b) the greatest energy of β^- particle you would expect to observe;
(c) the value of Q for this decay.

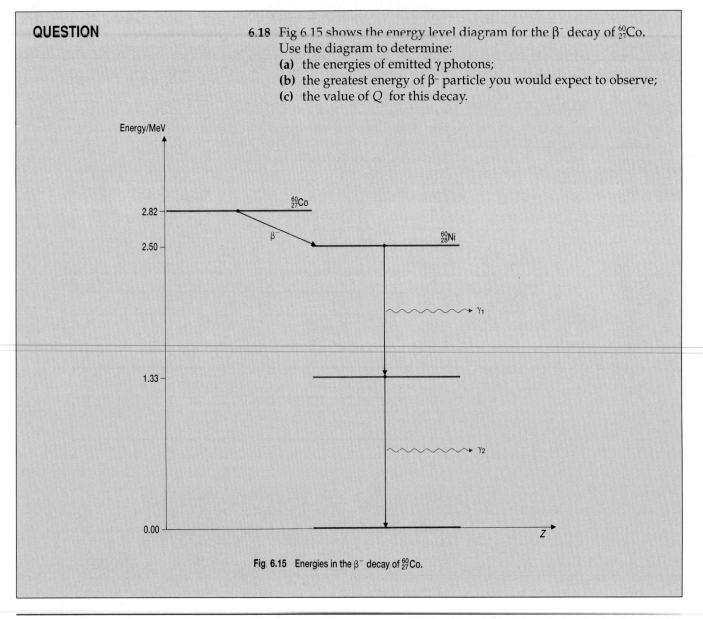

Fig. 6.15 Energies in the β^- decay of $^{60}_{27}Co$.

THE PROCESSES OF RADIOACTIVE DECAY

6.4 NEUTRINOS

The story of how the existence of neutrinos was first postulated, and then demonstrated experimentally more than 20 years later, is an interesting example of the way in which ideas from theoretical physics can guide experimenters in their work.

As we have seen in Section 6.2, the existence of the neutrino was postulated to explain the shape of the energy spectrum of β particles shown in Fig 6.6(a). Something was carrying off part of the energy released, and in 1932 Pauli suggested that there must be a new particle responsible. It was not until 1956 that direct evidence was established for the existence of the neutrino.

Neutrino means 'the little neutral one'. Its properties mean that it is very difficult to detect: it has zero charge, zero mass, and interacts very weakly with matter. Indeed, neutrinos are so weakly interacting with matter that they can pass straight through the Earth with very little chance of being absorbed. We are constantly bathed in a stream of neutrinos from the Sun, but very few of them are absorbed. They can pass through hundreds of light years' thickness of matter with only a 50 per cent chance of being absorbed. This is why they are so difficult to detect.

It is difficult to prove experimentally that a particle has zero mass. So far it has been reliably established from β energy spectra that the mass of the neutrino is certainly less than one two-thousandth of the electron mass.

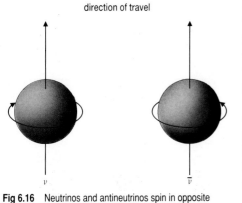

direction of travel

ν $\bar{\nu}$

Fig 6.16 Neutrinos and antineutrinos spin in opposite directions, relative to their direction of travel.

Neutrinos and antineutrinos

There are several types of neutrino associated with different nuclear interactions. The ones we are concerned with are the electron neutrino ν (associated with β⁺ emission), and the electron antineutrino ν̄ (associated with β⁻ emission). These we have simply referred to as the neutrino and antineutrino, but there are others.

Neutrinos belong to the family of particles called leptons, meaning light (weight) particles. Electrons and positrons are also leptons. When we said earlier, in Section 5.1, that the number of particles in a decay equation must be balanced, it would have been more accurate to say that the number of leptons must be conserved. Protons and neutrons are not leptons, and so do not count.

It is hard to imagine, for such nebulous particles, just what the difference could be between a neutrino and its antiparticle, the antineutrino. The difference is shown in Fig 6.16; one has left-handed spin, the other has right-handed spin.

Detecting neutrinos

Antineutrinos were not directly detected until the invention of the nuclear reactor provided a suitable source. The reaction used to detect the antineutrinos was:

$$p + \bar{\nu} \rightarrow n + e^-$$

This was shown to take place in tanks of water placed near the reactor. At the same time efforts were made to look at reactions involving the neutrino, rather than the antineutrino. The reaction looked for here was:

$$^{37}_{17}Cl + \nu \rightarrow {}^{37}_{18}Ar + e^-$$

The resulting isotope of argon was never detected, since nuclear reactors do not produce neutrinos. This showed that there was a genuine difference between the neutrino and the antineutrino.

helium in → | → helium + argon out

reactor

C_2Cl_4

Fig 6.17 A neutrino detector.

The technique for detecting neutrinos is shown in Fig 6.17. Chlorine is present in the form of tetrachloroethene; the apparatus is periodically flushed out with helium gas, and argon atoms sought for. Since they are produced at a rate of less than one per day, you will appreciate the sensitive techniques required to detect them!

Neutrinos from the stars

Neutrinos might seem to be scarcely worth worrying about since they interact so weakly with matter. However they do play an important part in some of the nuclear processes which go on in stars.

Two β^+ decays which occur in stars are:

$$^{13}_{7}N \rightarrow {}^{13}_{6}C + e^+ + \nu + \gamma$$

$$\text{and } {}^{15}_{8}O \rightarrow {}^{15}_{7}N + e^+ + \nu + \gamma$$

Both of these produce energetic neutrinos, and the Earth is constantly bathed in a flux of such neutrinos from the Sun. A detector like that shown in Fig 6.17 is housed 1500m underground in a mine in South Dakota (Fig 6.18). It contains over 600 tonnes of tetrachloroethene, and it has been able to show the passage of neutrinos through the earth.

There is a problem, though. The flux of neutrinos is only one-third of that expected. Clearly we do not know all that there is to know about nuclear processes in the Sun; there may also be something wrong with our understanding of the way the detector works.

Neutrinos from a supernova

In February 1987 a supernova was spotted in the southern skies by an observant Canadian astronomer, Ian Shelton from Toronto. A supernova (see Fig 6.19) occurs when an ageing star blows itself apart under the pressure of neutrinos produced in its interior. This supernova, named SN 1987a, was the first such astronomical event to be detected this century.

It was believed that, when a supernova occurs, a great burst of neutrinos would be released into space. Neutrino scientists rushed to their detectors when the news of the supernova was released. Sure enough, recorders at three different stations showed the passage of a burst of neutrinos lasting several seconds. The neutrinos arrived at the detectors several hours before the light from the supernova; confirmation that they travel through space

Fig 6.18 The neutrino detector in the Homestake Gold Mine, South Dakota.

Fig 6.19 A supernova, the Crab Nebula, whose explosion was observed in 1054 AD.

THE PROCESSES OF RADIOACTIVE DECAY

at the speed of light. (The light itself arrived after the neutrinos, because it interacted with interstellar dust as it travelled through space.)

SUMMARY

Alpha particles are trapped within the nucleus behind a Coulomb barrier; they may escape by the process of quantum mechanical tunnelling. Beta particles have a range of energies from 0 to Q. From this was inferred the existence of the neutrino. Experiments have since established the existence of these particles. Gamma photons show line spectra – a limited number of precise wavelengths. Gamma photons are emitted by excited nuclei after alpha or beta emission.

EXAMINATION QUESTIONS: Theme 2

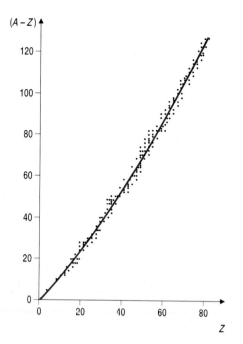

T2.1

(a) A radioactive isotope of strontium (Sr) has an atomic (proton) number Z of 38 and a mass (nucleon) number A of 90. It decays by the emission of a β⁻ particle to form an isotope of yttrium (Y).

   (i) What is the composition of the nucleus of the strontium isotope?

   (ii) What changes take place when the β particle is emitted? Write an equation representing the changes.

(b) In the diagram the value of (A–Z) has been plotted against the value of Z for a large number of nuclides. The line has been drawn approximately through the points obtained for stable nuclides.

   (i) What does the graph indicate about the number of neutrons present in the nuclei of stable nuclides which are light compared to those which are heavy?

   (ii) Assuming that a radioactive disintegration of an unstable nuclide results in a new nuclide nearer to the line for stable nuclides, state and explain whether you would expect the point for $^{90}_{38}$Sr to occur to the left or to the right of the line.

(AEB 1986)

T2.2

(a) Sketch a graph of the binding energy E per nucleon against the mass (nucleon) number A for the naturally occurring isotopes, indicating an approximate scale for A.

(b) Use your graph to explain briefly how energy may be obtained from fission of heavy nuclei.

(c) The following is a possible nuclear reaction:

$$^{3}_{1}H + ^{2}_{1}H \rightarrow ^{A}_{Z}X + ^{1}_{0}n + 17.6 \text{ MeV}$$

   (i) Name the nucleus X.

   (ii) Using the data given below, calculate the atomic mass of X in u.

      mass of neutron   = 1.0087 u
      mass of $^{2}_{1}$H atom = 2.0141 u
      mass of $^{3}_{1}$H atom = 3.0161 u
      1 u   = 931 MeV

(O&C 1987)

T2.3

(a) What is meant by the binding energy of a nucleus? Sketch a graph of binding energy per nucleon against nucleon number for the naturally occurring nuclides, indicating approximate scales on the axes of your graph.

(b) One form of the binding energy equation of the liquid drop model for a nucleus of nucleon number A, proton number Z and neutron number N is given below, where a, b, c and d are constants.

$$\text{Binding energy} = aA - bA^{2/3} - \frac{cZ^2}{A^{1/3}} - \frac{d\,(N-Z)^2}{R}$$

(i) State the origins of the terms in the equation containing b, c and d.

(ii) Explain how the equation is consistent with the general shape of the graph in (a), indicating which terms are more important for light nuclei and which for heavy nuclei.

(c) What are the main characteristics of the strong forces between nucleons in the nucleus?

(JMB 1986)

T2.4

(a) (i) Sketch a graph showing the potential energy of an α particle near a heavy nucleus as a function of their separation. State what is meant by the tunnel effect and use your statement to explain how a nucleus can decay by emitting an α particle.

(ii) Account for the fact that ^{226}Ra emits α particles of two distinct energies accompanied by γ radiation. Calculate the wavelength of the γ radiation if the energies of the α particles are 7.65×10^{-13} J and 7.36×10^{-13} J.

Speed of light = 3.00×10^{8} m s^{-1}

The Planck constant = 6.63×10^{-34} J s

(b) Sketch a typical kinetic energy spectrum for the β^{-} particles produced in radioactive decay. Explain the nature of this spectrum, and state how it is characteristic of the decaying nuclide.

(JMB 1984)

T2.5

(a) Describe the properties of the neutrino. How does the postulate of the existence of the neutrino account for the observed features of beta ray energy spectra?

(b) The nuclide ^{37}Ar can decay by electron capture.

(i) Explain what is meant by *electron capture*.

(ii) Calculate the energy released in the decay of ^{37}Ar, using data selected from the table on the next page.

(iii) State how the energy released is carried off following the decay.

In the following table, the mass of an isotope is given for a neutral atom of the substance and is quoted in unified atomic mass units, u.

1 u is equivalent to 931.3 MeV.

Name	Symbol	Charge number Z	Mass number A	Atomic mass M (u)
electron	$e^{\pm}$	± 1	–	0.000 55
proton	p	1	1	1.007 28
neutron	n	0	1	1.008 67
hydrogen	^{1}H	1	1	1.007 83
helium	^{4}He	2	4	4.002 60
chlorine	^{35}Cl	17	35	34.968 85
	^{36}Cl	17	36	35.968 31
	^{37}Cl	17	37	36.965 90
	^{38}Cl	17	38	37.968 00
argon	^{36}Ar	18	36	35.967 55
	^{37}Ar	18	37	36.966 77
	^{38}Ar	18	38	37.965 72
	^{39}Ar	18	39	38.964 32
	^{40}Ar	18	40	39.962 38
potassium	^{37}K	19	37	36.973 4
	^{38}K	19	38	37.969 1
	^{39}K	19	39	38.963 71
	^{40}K	19	40	39.964 01
	^{41}K	19	41	40.961 83

(JMB 1979)

Theme 3

NUCLEAR TECHNOLOGY

The first four decades of the twentieth century saw great advances in our knowledge of radioactivity and of the structure of the nucleus. You have studied these in the first two themes of this book. Since 1940 there have been great developments in nuclear technology; that is, in the exploitation of our scientific knowledge.

Nuclear technology includes the use of radiation in industry, medicine and other fields; nuclear power; and nuclear weapons. This technology was spurred on by the Second World War, and has been at the centre of controversy ever since. All new technologies have their supporters and their opponents. They have their benefits and their disadvantages. But no other technology this century has aroused such passions amongst scientists, engineers and the public.

Perhaps you can think of reasons why nuclear technology is so controversial – you may have strong ideas about this yourself. In these three chapters you can find out about the physics which lies behind our uses of nuclear materials. As you do so, you should consider whether your increasing knowledge changes or strengthens your opinions to any extent.

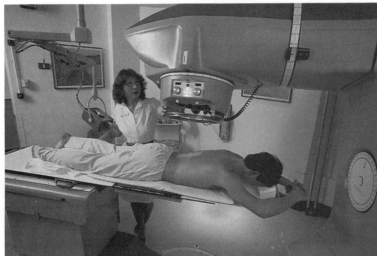

A cancer patient undergoing gamma radiotherapy.

Police patrol the perimeter of a base housing American nuclear weapons. Protestors have decorated the barbed wire fence.

Chapter 7

USING RADIOACTIVE MATERIALS

Radioactive materials are used in an increasing variety of applications. This chapter looks at some of these uses and relates them to the properties of radioactive materials. In order to use radioactive materials, it is necessary to be able to detect the radiations they produce. This chapter includes a brief survey of detectors. Finally, since radiation can be dangerous, there is a discussion of the health hazards involved in using radioactive materials, and of the safety precautions needed in industry and in school and college laboratories.

LEARNING OBJECTIVES

After studying this chapter you should be able to:

1. handle radioactive materials with appropriate care and attention to safety;

2. outline the effects of α, β and γ radiations on the body;

3. describe how α, β and γ radiations may be detected using photographic film, the ionisation chamber, the Geiger tube, and solid state detectors;

4. state the following properties of α, β, and γ radiations: charge, mass, penetration of air and solid materials;

5. relate the uses of radioisotopes to their properties.

---

radioactive metal foil

mounting pin

gauze window

Fig 7.1 **(a)** The construction of a typical radioactive source for use in school and college laboratories.

(b) Radioactive sources and their storage box.

Handling radioactive materials

Before you embark on any experiments involving radioactive sources, you need to know the rules for safe handling of the sources available to you. These rules are intended for your protection, and when you have completed your study of this chapter you should understand the reasons for them.

The sources which are available to you are **closed sources**. This means that the radioactive material is enclosed so that you cannot easily come into direct contact with it.

Fig 7.1 shows the construction of a typical source, in which the radioactive source is in the form of a metal foil and the radiation emerges through the open end (the window) of the holder. The picture also shows the lead-lined box in which such sources are stored.

In the discussion of the properties of radioactive materials in the next section you should think about just how safe and secure your radioactive sources are.

Laboratory rules

- When using radioactive sources always follow the instructions of your teacher.

- Always handle sources using forceps; never use your bare hands.

- Hold sources so that the open window is pointing away from you and your neighbour. Don't try to look into the source.

- Work well away from other students in the laboratory.

Once you have placed the source in a suitable fixed holder, it is usually unnecessary to stand near it during the course of an experiment.

7.1 PROPERTIES OF RADIATIONS

Charge, mass and speed

We have already discussed the nature of alpha, beta and gamma radiations; the signs of their charges may be demonstrated by their deflections in a magnetic field.

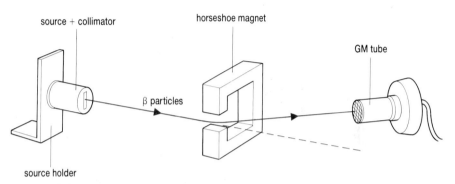

Fig 7.2 An experiment to demonstrate the deflection of β particles in a magnetic field.

Fig 7.2 shows how this may be investigated in the laboratory. The deflection of particles in a magnetic field depends on three things: their charge, their mass and their speed. Fleming's left-hand rule will tell you the direction of the force experienced by the particles. (A more elaborate experiment is needed to determine the magnitudes of charge, mass and speed.) In practice it is not possible to obtain sufficiently strong magnetic fields in the school or college laboratory to detect the deflection of alpha particles.

QUESTIONS

These questions relate to Fig 7.2.

7.1 Use Fleming's left-hand rule to determine which pole of the magnet (upper or lower) is the north pole.

7.2 Some β particles are deflected more than others in the magnetic field. What does this tell you about the speeds of these particles? Which would you expect to be deflected least?

Penetration by radiation

In order to know how to protect ourselves from hazardous radiation it is necessary to have some idea about the ability of different radiations to penetrate matter. Fig 7.3 shows how to investigate this.

α particles may be absorbed by a thin sheet of paper. The cloud chamber photographs of Fig 6.1 show that α particles are absorbed by a few centimetres of air. Paper is about 1000 times denser than air, and so a fraction of a millimetre of paper is enough to absorb the energy of the α particles.

As α particles travel through air, they collide with oxygen and nitrogen

Fig 7.3 Investigating the absorption of (a) α, (b) β and (c) γ radiations.

molecules. In each collision, they lose some energy in ionising the air molecule. After perhaps 100 000 such collisions, they have lost all of their energy and they are absorbed. In paper, the molecules are much closer together and so the penetration of paper by α particles is much less than their penetration of air.

β particles are more penetrating than α particles. They travel very fast and so they interact less readily with the atoms and molecules of the matter through which they are passing. A typical track from a cloud chamber is shown in Fig 7.4.

Most of the β particles from the sources used in schools are absorbed by 3 or 4 mm of aluminium.

γ rays are the most penetrating of all. In fact, because of the way in which γ rays interact with matter, it is not possible to say that a particular thickness of material is enough to absorb all rays. Rather, we have to talk about the **half-thickness** of a material. This is the thickness of material which is enough to absorb half of the γ rays falling on it. If the thickness of the material is doubled, half of the remaining rays are absorbed; a quarter remain unabsorbed. Three times the original thickness will leave an eighth of the rays unabsorbed, and so on.

Fig 7.4 The tracks of β particles in a cloud chamber.

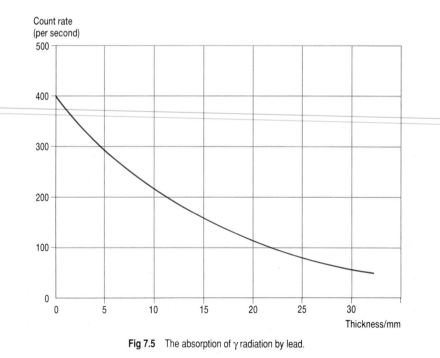

Fig 7.5 The absorption of γ radiation by lead.

USING RADIOACTIVE MATERIALS

Fig 7.5 shows how the intensity of γ rays decreases as they pass through lead. This graph should remind you of the way in which a radioactive material decays; it shows an exponential decrease.

γ rays are absorbed by the nuclei of lead atoms. A single γ photon has a 50 per cent chance of being absorbed by a nucleus within one half-thickness of lead. If it is not absorbed within this distance it then has a 50 per cent chance of being absorbed by a nucleus in the next half-thickness, and so on.

QUESTIONS

7.3 Compare the cloud chamber traces of Figs 6.1 and 7.4. What feature shows that β particles interact less strongly with air than do α particles?

7.4 Use Fig 7.5 to estimate the half-thickness of lead to γ rays.

7.5 Nuclear reactors are very strong sources of γ rays. They are often contained in a concrete shell, 2 m thick. A 1 m thickness of concrete reduces the intensity of radiation by a factor of 1000. What fraction can penetrate a 2 m thick shell?

INVESTIGATION

Gamma rays in air

Because air is much less dense than lead, few γ rays are absorbed by air. They travel outwards in straight lines, like rays of light from a lamp. Their intensity decreases with distance as they spread out. You can investigate this spreading out, which obeys an inverse square law, as shown in Fig 7.6.

Fig 7.6   Investigating the inverse square law for γ radiation.

Because γ rays are very penetrating, they emerge from all sides of the source. They do not just emerge from the open window. You should devise an experiment to investigate this. Try to answer these questions: Is the radiation equally intense from all sides of the source? At what distance from the source is the intensity of γ radiation comparable with the level of background radiation?

Safe practice

By now you should have an understanding of the properties of radiation which will allow you to appreciate the rules for handling sources in the laboratory.

There is need for caution when handling radioactive sources, but there is no need for alarm. Remember, the sources available to you are designed for use by students under supervision, and your teacher will be aware of procedures for their safe handling.

α, β and γ radiations can all damage living tissue. You need to avoid exposing yourself to this hazard. α particles are readily absorbed. They cannot penetrate even the outer layer of dead cells of your skin. However, you would not have this protection from an α source inside the body. Hence the importance of avoiding direct contact with sources.

β particles and γ rays have less damaging effects than α particles of the same energy. This is because, as we have seen, α particles interact most strongly with the matter through which they are passing. However, it is important to minimise your exposure to all radiations, and you should follow the rules above.

If you have carried out the investigation you will have found out the distance you have to be from a γ source for the radiation level to be comparable to the background level. Of course, this depends on the strength of the source and the background level; typically, this distance is about a metre. Hence, if for most of the time when you are doing an experiment you keep your body at least one metre away from the source, you will avoid exposing yourself to any unnecessary doses of radiation.

7.2 DETECTING RADIATION

Since α, β and γ radiations are invisible to us, we need suitably designed detectors to show their presence. What follows in this section is a summary of how the most common detectors work; you can find more details in a standard physics textbook (see Appendix A).

Detectors might be expected to tell us two things: the nature of the radiation being detected, and the energies present.

Fig 7.7 Film badges, used to monitor exposure to radiation.

USING RADIOACTIVE MATERIALS

Fig 7.8   A spark detector.

Fig 7.9   An ionisation chamber; the anode cathode voltage pulls apart the ion pairs formed by incoming radiation.

Photographic film

Radioactivity was originally discovered by accident, by Becquerel in 1896. He stored some photographic plates in a drawer with some uranium salt samples. When he developed the plates, he found that they were fogged. He deduced that the uranium salt was producing some kind of invisible radiation, and the study of radioactivity was born.

Fig 7.7 shows various **film badges** of the kind used to monitor the exposure to radiation of workers in industry and in scientific laboratories, who are likely to be exposed to doses of ionising radiations. (These include X-rays, as well as α, β and γ radiations.) Each consists of a small photographic film mounted in a hinged plastic holder. The holder has plastic and metal windows. When the film is developed, the pattern of exposure can be interpreted in terms of β and γ radiations and their energies.

Ionisation in gases

Several types of detector use the fact that radiation causes the ionisation of gases. The simplest of these is the **spark detector** (Fig 7.8). A fine wire is held at a high voltage relative to a metal plate or gauze. When an α source is brought close to the wire the air is ionised and becomes conducting. Sparks jump between the wire and the plate. This is a very simple way of demonstrating the detection of α particles, although it is not quantitative.

Other ionisation detectors are elaborations of the simple spark detector. The **ionisation chamber** shown in Fig 7.9 is like a 'rolled-up' spark counter. The anode is the central rod; the cathode is the outer case. A voltage of, perhaps, 50 V is maintained between them.

Radiation enters through the open end, or through the walls. Air molecules are ionised by the radiation, and are pulled apart by the anode–cathode voltage. Negative ions move to the anode, positive ions to the cathode. In effect, a current flows through the air in the chamber. This may be detected by amplifying the current in the external circuit.

QUESTIONS

7.6   Some types of radiation are more easily detected with an ionisation chamber than others. Which type of radiation (α, β or γ) would you expect to produce the most ion pairs per particle entering the chamber?

7.7   If each α particle entering the chamber produced 10^4 ion pairs, and one α particle arrived per second, what current would you expect to flow in the external circuit? Would you expect this to be measurable?

7.8   If the smallest current which you can measure is $1\,\text{pA}$ (10^{-12} A), what is the smallest number of α particles per second which you can detect using an ionisation chamber?

The Geiger counter

The **Geiger–Müller tube** uses gas multiplication to produce a bigger pulse of current for each detected particle of radiation. It is a sealed tube, similar to an ionisation chamber (but smaller), filled with gas. This gas is argon, at low pressure, mixed with bromine. (See Fig 7.10.)

When an argon atom is ionised by an incoming particle, the electron produced is accelerated towards the central anode. (The bromine gas is present in the tube as a quenching agent; its purpose is to absorb the energy of the positive argon ions as they accelerate towards the cathode.) This electron collides with other argon atoms, producing a cascade of electrons. One ionisation event may result in 10^8 or more electrons reaching

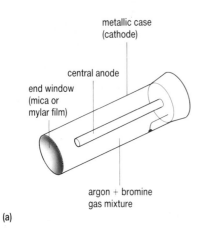

metallic case
(cathode)

central anode

end window
(mica or
mylar film)

argon + bromine
gas mixture

(a)

Fig 7.10   A Geiger-Muller tube; a single ionisation event results in a cascade of electrons reaching the anode.

the anode. This gives a measurable pulse in the external circuit. Pulses may be counted by an electronic counter, or they may be used to give a display on the scale of a ratemeter.

Thus a Geiger counter can be used to detect individual particles arriving at the tube, whereas the ionisation chamber can only give an average reading for relatively intense radiations.

Solid state detectors

Solid state detectors are based on **p-n junction diodes**. (These are the familiar semiconductor diodes used for rectifying alternating current.) Such a diode is connected up to a power supply so that it is reverse biased, and on the point of conducting (Fig 7.11). Incoming radiation produces ions in the semiconductor material, and the extra free electrons are enough to make the diode conduct. A pulse of current flows around the circuit for each particle detected. The size of the pulse shows the type of radiation present.

Other detectors use substances which **fluoresce** (emit flashes of light) when struck by ionising radiation (see Table 7.1). Rutherford used zinc sulphide to detect α particles in his early experiments. The tiny flashes of light produced may be amplified using a light sensitive cathode, which emits an electron into a photomultiplier tube, resulting in a cascade of electrons which is then readily detected.

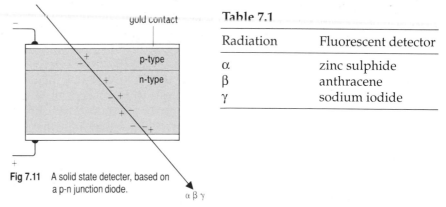

Fig 7.11   A solid state detecter, based on a p-n junction diode.

Table 7.1

Radiation	Fluorescent detector
α	zinc sulphide
β	anthracene
γ	sodium iodide

7.3 USING RADIOISOTOPES

In this section you will find brief descriptions of several uses of radioisotopes (Figs 7.12–7.24), which you should be able to understand from your knowledge of the properties of radioactive materials.

Fig 7.12   This heart pacemaker has a nuclear powered battery. A small quantity of plutonium decays by α emission; the heat generated is converted to electricity by a semiconductor thermopile. The battery is 35mm long.

Fig 7.13   Radioactive iridum strips are implanted in tumorous tissue. The radiation destroys the cancerous cells.

Fig 7.14   A gamma camera detects the position of a radioactive tracer in the patient's bloodstream. In this case, a scan of the heart is produced.

Fig 7.15   A gamma knife concentrates 200 individually harmless gamma rays at a precise point in the patient's brain.

Fig 7.17   The thickness of metal tyre cord is monitored at the Avon Rubber plant at Melksham; the thickness gauge uses a strontium-90 source.

Fig 7.16   The outflow of dirty water from a china clay works in Cornwall is monitored using radioactive tracers.

Fig 7.18   In industrial radiography, solid objects are inspected to find any flaws, in a similar way to a medical X-ray examination. Here, the Rolls Royce RB-211 engines of a British Airways 747 jet are being inspected.

USING RADIOACTIVE MATERIALS

Fig 7.19 Here, sensitive photographic film is wrapped around a part under investigation.

Fig 7.22 Food may be irradiated to kill parasites, bacteria and pest insects, and to inhibit germination, sprouting or premature ripening. This can help to prolong the shelf-life of the food. Some irradiated foods contain unpleasant-tasting biochemicals, rendering them unpalatable. The food itself does not become radioactive.

Fig 7.20 Medical equipment is sterilised using gamma radiation from a powerful cobalt-60 source. Boxes of equipment are being loaded on to the conveyor belt before entering the irradiation plant.

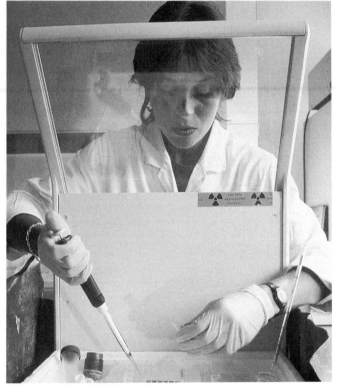

Fig 7.23 Biochemicals may be "labelled" with radioactive atoms. Their chemical nature is unchanged, and they are therefore useful for tracing chemical processes in living organisms. This scientist is using radioactive tracers in research in immunology.

Fig 7.21 Geologists find the age of rocks from the relative proportions of decayed and undecayed isotopes. Here, a member of the British Antarctic Survey collects samples.

Fig 7.24 Radiocarbon dating, in which the ratio of carbon-14 to carbon-12 is determined, allows archaeologists to find the age of dead materials, such as this mummified cat from Egypt.

USING RADIOACTIVE MATERIALS

Some applications depend on the energy released in nuclear decay; others on the penetrating power of different radiations; others on the exponential decay of radioactive materials with time; others on the detectability of low levels of radiation; still others on the effects of ionising radiations on biological materials. Which are which?

Where do these useful materials come from? The first radioisotopes available were found in naturally occurring minerals. However, these represented a limited range, and nowadays many more are produced by irradiating selected elements in nuclear reactors. Here they absorb neutrons and become unstable. Other isotopes are produced using particle accelerators, or by extraction from spent nuclear fuel.

Nuclear power and nuclear weapons, which consume large amounts of uranium, are discussed in Chapter 8.

ASSIGNMENT

Read about the various uses of radioisotopes shown in Figs 7.12 to 7.24, and answer the following questions.

7.9   Which applications depend on the detectability of small amounts of radioactive material?

7.10 Which depend on the penetrating power of radiation, particularly γ radiation?

7.11 Which depend on knowledge of the half-lives of radioisotopes?

7.12 Which rely on the release of energy in radioactive decay?

7.13 Which depend on the effect of radiation on living materials?

7.14 For which applications are only weak sources needed? Which need strong sources?

7.4 EFFECTS OF RADIATION ON THE BODY

The effects of radiation on the body are described in different ways. These effects depend on the nature of the radiation and the energy it gives to the body. In this section you can find out about some of these effects and the ways in which they are quantified.

Absorbed dose

This quantity describes the energy transferred to a body, per kilogram of tissue. The unit of absorbed dose is the gray, symbol Gy. One gray is equal to one joule of energy absorbed per kilogram. (If you catch a fast-moving cricket ball it will give you about one joule of energy for each kilogram of your mass. This should give you some idea of the magnitude of the gray.)

Dose equivalent

You already know that α particles interact more strongly with matter, and have a greater ionising effect than β particles or γ rays. How can we take account of this?

We have to multiply up the dose in grays by a **quality factor**, which accounts for the relative effectivenesses of different radiations. Table 7.2 lists the current values of quality factor for some types of radiation.

(Quality factors are determined from experimental evidence of the effects of radiation on living tissue. They are revised periodically as more information becomes available.)

From the table you will see that a given absorbed dose of α particles is 20 times as effective in damaging tissue as an equal dose of β particles or γ rays. We can calculate the **dose equivalent** of radiation using the equation:

dose equivalent = absorbed dose × quality factor

The unit of dose equivalent is the sievert, symbol Sv.

Table 7.2

Radiation	Quality factor
α	20
β	1
γ	1
slow neutrons	2.3
fast neutrons	10

Units old and new

The gray and the sievert are the SI units of absorbed dose and dose equivalent. However, you are likely to come across older, non-SI units of these quantities, the rad and the rem. Table 7.3 shows how these are related.

Table 7.3

Quantity	SI unit	Old unit	Conversion
absorbed dose	gray, Gy	rad	1 rad = 0.01 Gy
dose equivalent	sievert, Sv	rem	1 rem = 0.01 Sv

Radiation and the body

We all know that radiation can damage living tissue. It may damage the genetic material of cells, so that their ability to reproduce themselves is defective. This may result in cancer in an individual, or in genetic disorders in offspring. Alternatively, it may so disrupt cells that they die, and the body has difficulty in repairing them.

The information we have about the effects of radiation comes from two sources: firstly, investigation of the survivors of the bombs dropped on Hiroshima and Nagasaki, patients who have been strongly irradiated for medical purposes, and workers who have been exposed to high levels of radiation in uranium and other mines, can tell us about the link between radiation exposure and human cancer. Secondly, information about hereditary defects produced by irradiation comes from animal experiments. Similar effects may occur in humans, but no significant defects have yet been found which could be attributed to irradiation.

We can classify these effects as **stochastic** and **non-stochastic**. In general, if an effect is stochastic the probability of its occurring increases with increasing dose. (Stochastic means random.) There is no minimum dose required. We can contract cancer from a very small dose of radiation, although a larger dose means a greater risk. Typically, stochastic effects appear years after the exposure which causes them.

Non-stochastic effects include radiation burns and radiation sickness. These occur typically above a certain threshold dose.

Table 7.4 shows the threshold dose equivalents for various non-stochastic effects of radiation on the human body. Of course, it is not possible to draw up such a table for stochastic effects.

Table 7.4 Effects of radiation on the human body.

Dose (Sv)	Effect	Dose (rem)
20	radiation sickness due to damage to central nervous system	2000
10	radiation burns: blistering of skin	1000
5	radiation sickness: gastro-intestinal	500
5	damage to eyesight	500
3	radiation burns to skin	300
2	radiation sickness: bone marrow	200
1	temporary sterility (women)	100
0.5	blood count reduced	50
0.1	temporary sterility (men)	10

Radiation doses in the UK

To put the figures in Table 7.4 in perspective they should be compared with the typical doses we receive in everyday life. The average annual dose received by citizens in the UK is about 2mSv. Note that this is about one fiftieth of the lowest dose quoted in Table 7.4.

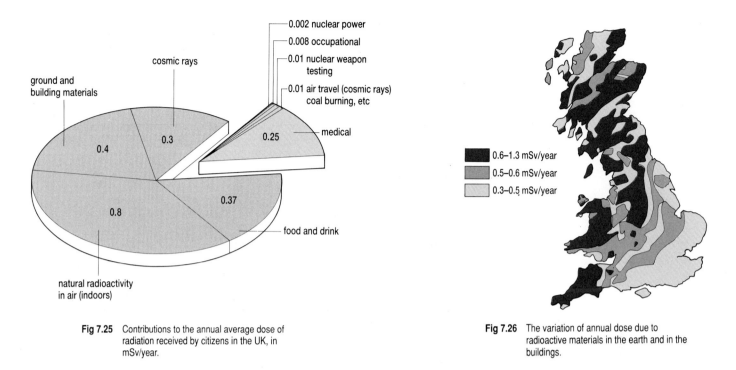

Fig 7.25 Contributions to the annual average dose of radiation received by citizens in the UK, in mSv/year.

Fig 7.26 The variation of annual dose due to radioactive materials in the earth and in the buildings.

Fig 7.25 shows the contributions to this average annual dose from different sources. Fig 7.26 shows how the background radiation due to the ground and buildings varies within the UK.

QUESTIONS

These questions relate to the data in Figs 7.25 and 7.26.

7.15 Which sources of radiation are of natural origin and which are of artificial origin? What proportion of the average dose comes from artificial sources?

7.16 People working with radiation are allowed a maximum annual dose of 50 mSv. By how many times does this exceed the annual average dose?

7.17 In the year following the Chernobyl accident the average dose received was increased by about 0.07 mSv. By what percentage did this increase the average annual dose?

7.18 Which contributions do you think may have changed significantly over the last twenty years, and in what way?

7.19 Individual doses vary according to location, occupation and lifestyle. If you wanted to reduce the dose which you currently receive how might you go about it?

SUMMARY

Alpha particles are readily absorbed by thin paper or by a few centimetres of air. Beta particles are absorbed by a few mm of aluminium. Gamma rays are absorbed differently; their intensity decreases exponentially as they pass through materials. About 1 cm of lead will absorb half the gamma rays.

Alpha emitters are dangerous if ingested into the body. Gamma sources are the most hazardous outside the body. An understanding of the properties of radiations leads to an awareness of the rules for safe handling radioactive materials.

The detection of radiation is based upon the ionisation of air, other gases, or semiconductors.

There is a great range of applications of radioisotopes – in industry, the health service, etc. These depend on the penetrating power of radiation, its effect on living material, the detectability of low levels of radioisotopes, and their exponential decay.

USING RADIOACTIVE MATERIALS

Chapter 8

NUCLEAR FISSION REACTORS

In terms of the volume of material used, by far the greatest uses of radioisotopes are in the generation of power in nuclear reactors, and in nuclear weapons. These two technologies have grown up together during the last fifty years. In this chapter you will find out about the physics and technology of fission reactors, and the similarities and differences between fission reactors and fission weapons.

LEARNING OBJECTIVES

After studying this chapter you should be able to:

1. describe the neutron-induced fission of a massive nucleus;

2. calculate the energy released in a fission event;

3. describe how a chain reaction may occur, and use correctly the terms critical reaction and critical mass;

4. explain the role of a moderator and control rods in a practical reactor;

5. compare the functioning of a fission reactor and a fission bomb;

6. outline the problems, including hazards, associated with the handling of nuclear fuels and nuclear waste materials;

7. describe the similarities and differences of construction and operation of thermal and fast nuclear reactors.

8.1 THE PHYSICS OF FISSION

The discovery of nuclear fission

In the 1930s physicists and chemists attempted to make **transuranic elements**; that is, elements with proton number greater than that of uranium, $Z = 92$. These had not been found in nature. To create these new elements the experimenters bombarded an isotope of uranium with neutrons. They hoped that uranium nuclei would capture neutrons and then, since they would be neutron-rich, they might decay by β emission. The reactions they were looking for included:

$$^{238}_{92}U + n \rightarrow {}^{239}_{92}U + \gamma$$

$$^{239}_{92}U \rightarrow {}^{239}_{93}Np + \beta^- + \bar{\nu}$$

$$^{239}_{93}Np \rightarrow {}^{239}_{94}Pu + \beta^- + \bar{\nu}$$

Indeed, the two transuranic elements $^{239}_{93}Np$ and $^{239}_{94}Pu$ were produced in this way. However, the experimenters also found that they had produced other, lighter nuclides, including isotopes of $_{56}Ba$ and $_{36}Kr$. Where might these have come from?

You will notice that the proton numbers of these two light nuclides, 56 and 36, add to give the proton number of uranium, 92. The light nuclides arise from the splitting, or fission, of uranium nuclei. A massive neutron-rich nucleus may decay by β⁻ emission or alternatively it may split in two. This was the first observation of **nuclear fission**.

Since the uranium nucleus became unstable and split after absorbing a neutron this is referred to as **neutron-induced fission**. Nuclei may split without first capturing a neutron; this is called **spontaneous fission**.

Fission products

A nucleus may split in two in many different ways. The daughter nuclei produced are always found to be in the middle range of the Periodic Table. Fig 8.1 shows the yields of fission products resulting from the fission of $^{235}_{92}$U. Notice that the vertical scale is logarithmic.

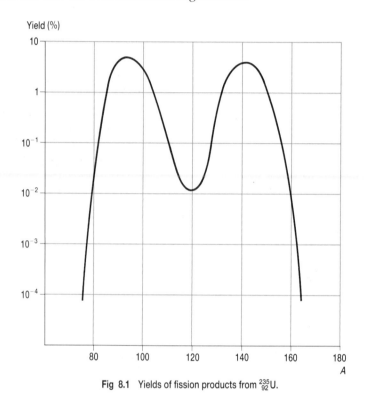

Fig 8.1 Yields of fission products from $^{235}_{92}$U.

QUESTIONS	Study the graph of Fig 8.1 and answer the following questions.

8.1 What are the mass numbers of the commonest fission fragments? How do these relate to the mass number of the parent nucleus?

8.2 Approximately what fraction of parent nuclei split in half, to give two nuclides having $A = 118$?

8.3 Do you conclude that massive nuclei tend to split into two equal or unequal parts?

ASSIGNMENT	You will recall from Chapter 3 that massive nuclei have a higher proportion of neutrons than do lighter nuclei. This helps to dilute the mutual repulsion of the protons. What happens to the numbers of neutrons in nuclear fission? A common reaction in a nuclear reactor is for a nucleus of uranium ($^{235}_{92}$U) to capture a single neutron. The resulting nucleus then splits. A typical pair of product nuclei is $^{141}_{56}$Ba and $^{92}_{36}$Kr.

NUCLEAR FISSION REACTORS

8.4 Write down two equations, one to represent the capture of a neutron by $^{235}_{92}\text{U}$, and the second to represent the subsequent fission event. Make sure that the equations balance. How many neutrons must be released in the course of the fission?

8.5 Would you expect any other particles, such as neutrinos, to be released?

Energy is also released in the course of nuclear fission. You can calculate the energy released in the fission event for which you have written equations above, using the information about particle masses given in Table 8.1.

8.6 Calculate the energy released (in MeV) in the fission event described above.

8.7 How does this energy compare with that released when an α particle is released in nuclear decay?

8.8 What forms would you expect this energy to take?

Table 8.1

Particle	Mass/u
$^{1}_{0}\text{n}$	1.008 665
$^{235}_{92}\text{U}$	235.043 93
$^{92}_{36}\text{Kr}$	91.926 25
$^{141}_{56}\text{Ba}$	140.914 34

The mechanism of nuclear fission

We can picture the process of nuclear fission in a similar way to alpha emission. Fig 8.2 shows the process in stages.

(a) A massive nucleus which is neutron-rich is unstable. It is held together by the strong nuclear force between nucleons.

(b) The energetic nucleus distorts. If it becomes sufficiently distorted the Coulomb repulsion between the protons in the two parts may be strong enough to separate them.

(c) Two highly excited fission product nuclei are formed.

(d) The product nuclei become more stable by emitting neutrons.

The neutrons emitted in this way are known as **prompt neutrons** because they are emitted promptly at the time of fission. The number produced is usually two or three per nucleus.

Other neutrons may be emitted by the fission products at a slightly later stage. This happens if a product nucleus decays to form a daughter nucleus in a very energetic state. It can become more stable by emitting a neutron; such a neutron is referred to as a **delayed neutron**, since it appears some time after the initial fission. Fig 8.4 shows a decay chain which involves the production of a delayed neutron.

(a)

(b)

(c)

(d)

Fig 8.2 A massive, neutron-rich nucleus undergoes fission.

INVESTIGATION

You can carry out a simple investigation of the fission of water drops. Remember that the liquid drop model is very successful in describing many features of the behaviour of nuclei.

- Take a flat plate about 20 cm across and smear 5 ml of cooking oil all over it. Put a few small drops of water on the plate (see Fig 8.3) and tilt the plate about. Observe the behaviour of the drops as they move slowly about. Do they distort or break up?

- Now replace the small drops with one larger drop, made from 10 or so smaller drops. Tilt the plate again. How does this larger drop behave?

- Sketch how the shape of the drop changes as it distorts as it moves around the plate. Does the drop break up into smaller drops? If the drop splits in two, are the two parts the same size? Are any very small droplets formed?

- How does the behaviour of these water drops compare with the behaviour of an unstable nucleus? Do you consider this a good model for nuclear fission?

Fig 8.3   Drops of water on a plate are a model for the process of nuclear fission.

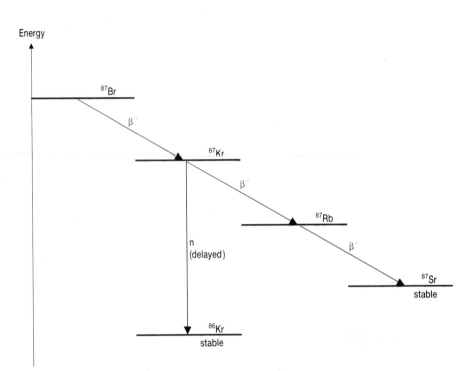

Fig 8.4   A decay chain, showing the production of delayed neutrons.

QUESTIONS

Think about the mechanism of the nuclear fission process and the liquid drop model of the nucleus. The model can help you to understand the mechanism.

8.9   As the energetic nucleus distorts (Fig 8.2(b)), what happens to its surface area? What happens to the surface energy term in the semi-empirical binding energy equation (see page 48)?

8.10 As the two product nuclei separate, what force pushes them apart? Is there any comparable force between water droplets formed when a larger drop breaks up, as in the investigation above?

Fission and binding energy

In Chapter 5, we saw that nuclei decay because this allows them to achieve a lower energy state. The binding energy per nucleon is greater after decay than before and some energy is released, together with a particle or particles. We can use the same ideas to understand why nuclear fission occurs.

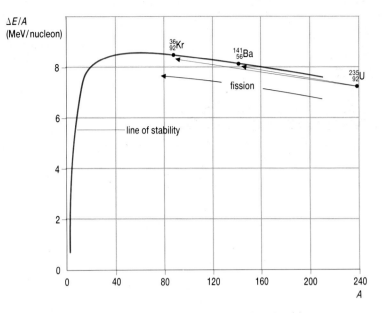

Fig 8.5 Nuclear fission results in more tightly bound nuclei.

Fig 8.5 is the graph of binding energy per nucleon against mass number, which you examined earlier in Chapter 4. The greatest value of $\Delta E/A$ occurs when A is approximately 80. Massive nuclei ($A > 200$) have significantly lower binding energy per nucleon. They can increase their value of $\Delta E/A$ and hence become more stable by splitting to form two lighter nuclei. The graph shows the positions of the nucleus $^{235}_{92}\text{U}$ and the fission products discussed above.

8.2 CHAIN REACTIONS

The energy released in the fission of a single nucleus of uranium is usually about 200 MeV. This is much greater than the energy released in alpha decay – typically 5 MeV. It is vastly greater than the energy released when an interatomic bond is broken in the conventional combustion of fuels such as coal and oil, which is only a few eV. Clearly the fission of uranium has great potential for energy production. In the rest of this chapter we will look at the ways in which this energy may be released, and the problems involved in controlling nuclear fission.

Chain reactions and chain letters
The principle of a chain reaction is very simple and probably familiar to you. To establish a chain reaction in uranium, and to control it, requires advanced technology. We have seen that when a massive nucleus, such as $^{235}_{92}\text{U}$, captures a neutron it splits and releases two or three more neutrons. In principle these fission neutrons could all go on to initiate other fission events, which would liberate more neutrons, and so on. The number of neutrons escalates rapidly and a chain reaction is established.

It is rather like a chain letter. You receive a letter which instructs you to send six letters to friends, who must all send six letters to friends of theirs. Sometimes these letters claim to have been going on for several years and you are asked to help break the world record. If such letters had been going around for years, and everyone had followed the instructions, you can imagine that the postal services would have collapsed under the burden! So what goes wrong?

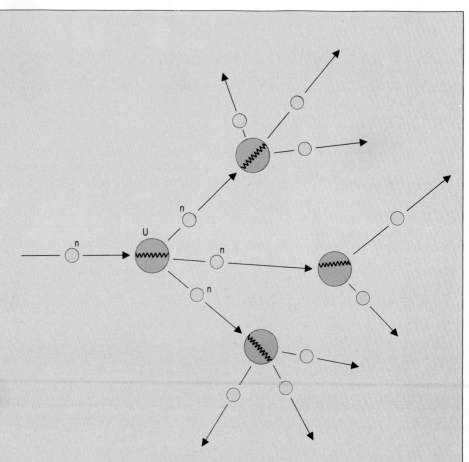

Fig 8.6 An idealised chain reaction.

QUESTIONS

Fig 8.6 shows, schematically, how a chain reaction occurs.

8.11 In the figure all the neutrons produced in the first fission event are depicted initiating further fissions. What might prevent this from happening? How many of the neutrons must go on to cause further fissions, if the chain reaction is to be maintained?

8.12 In the 'chain-letter' analogy, what stops the postal services from being overwhelmed?

A critical condition

The neutrons which result from the fission of a uranium nucleus have kinetic energies of several MeV. They are moving very fast, at speeds approaching the speed of light. They interact only weakly with other uranium nuclei as they pass by at high speed. There is a good chance that they will leave the piece of uranium, if it is small, before they have the chance to cause a further fission.

Thus there are two possibilities for a chain reaction: it may escalate out of hand, or it may die out as neutrons leave the material. For a self-sustaining chain reaction we need a balance between the rate at which neutrons are produced by fissions and the rate at which they are lost by causing fissions, by absorption, or by leaving the material.

When this balance is achieved, and the chain reaction is self-sustaining at a steady rate, we say that the reaction is **critical.** If the number of neutrons increases beyond the critical level we say that the reaction is supercritical.

Since neutrons which leave the material are lost to the reaction, it follows that a small mass of uranium is less likely to sustain a critical reaction than a large mass. This is because more atoms are near the surface in a small

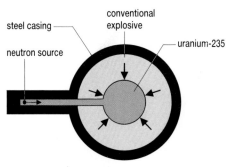

steel casing
conventional explosive
neutron source
uranium-235

Fig 8.7 The construction of a fission nuclear bomb.

mass, and more neutrons are therefore produced near the surface and easily lost.

There is a minimum mass needed, the **critical mass**, if a critical reaction is to be established at all. For uranium, a sphere of mass 15 kg (radius 6 cm, about the size of a grapefruit) is just large enough to be a critical mass.

An atomic bomb

There is a second way in which a critical reaction can be achieved. This is by strongly compressing a lump of uranium so that the uranium nuclei are squeezed together. This is made use of in some atomic bombs (Fig 8.7).

This bomb has three important parts: the fissile uranium; some conventional explosive, which is detonated to compress the uranium; and a strong container. This serves to trap the neutrons within the bomb and to hold the material together long enough for all of the uranium to be consumed.

QUESTIONS

detonator
explosive
cannon
uranium-235

Fig 8.8 A cannon-type nuclear bomb.

8.13 Explain why the critical mass for a long, thin cylinder of uranium is greater than for a sphere.

8.14 Fig 8.8 illustrates the principle of the cannon-type atomic bomb which was dropped on Hiroshima in 1945. It contained a total of 40 kg of uranium, more than twice the critical mass. Explain what happened to produce a nuclear explosion when the detonator set off the conventional explosive.

Fig 8.9 The devastating explosion of a nuclear bomb. This shows an American test explosion at Bikini Atoll in July 1946.

8.3 CONTROLLING CHAIN REACTIONS

In a fission nuclear bomb a catastrophic nuclear reaction is set off and a devastating release of energy occurs (see Fig 8.9). The explosion sends out a powerful pulse of electromagnetic radiation, together with a violent shock wave, a ball of fire and a giant cloud of radioactive smoke and dust. The entire chain reaction occurs in a fraction of a microsecond.

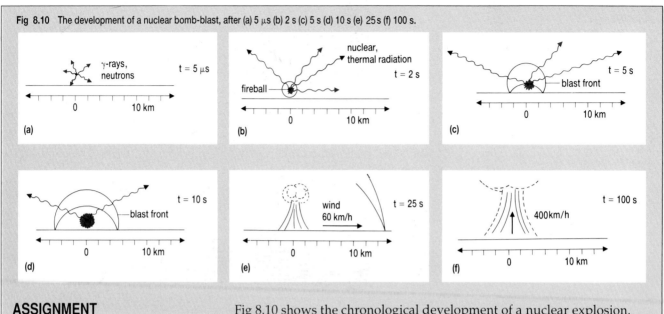

Fig 8.10 The development of a nuclear bomb-blast, after (a) 5 μs (b) 2 s (c) 5 s (d) 10 s (e) 25 s (f) 100 s.

ASSIGNMENT

Fig 8.10 shows the chronological development of a nuclear explosion. This shows the effects of an air-burst of a one megaton bomb. The energy released is the equivalent of a one-million ton conventional explosive bomb.

Read an account of the effects of nuclear weapons; some references are included in Appendix A. Write a brief description of the effect of a nuclear explosion. In what different ways would people close to and at a distance from the blast be affected?

Fission for electrical power

To make practical, non-destructive use of the energy available from fissile materials it is obviously necessary to be able to control chain reactions and to be sure that, in any reactor, the reaction cannot go out of control to the point where a nuclear power station becomes a nuclear bomb.

Before we go on to discuss how this is achieved, look at Fig 8.11. This compares the basic components of a nuclear power station and a conventional (coal- or oil-fired) power station. It is easy to identify the similarities and differences between these. You should understand the purpose of each component.

A chain reaction in uranium

Fig 8.12 shows what happens to the neutrons produced in a fission reaction. As we have seen, for a chain reaction to be established an average of one neutron from each fission event must go on to cause another such event.

Natural uranium consists mainly of two isotopes; the more important fissile isotope is $^{235}_{92}U$. Unfortunately for nuclear technologists, this isotope accounts for only 0.7 per cent of natural uranium. The fission reaction can be summarised as:

$$^{235}_{92}U + ^{1}_{0}n \rightarrow \text{fission fragments} + 2 \text{ or } 3 \text{ neutrons} + \text{energy}$$

Fig 8.12 What happens to neutrons produced in a fission reaction.

(a)

(b)

Fig 8.11 The principal components of power stations (a) coal-fired (b) nuclear. Note the similarities, and the differences.

The remaining 99.3 per cent is $^{238}_{92}$U. This second isotope can capture neutrons but it does not always undergo fission.

$$^{238}_{92}\text{U} + ^{1}_{0}\text{n} \rightarrow ^{239}_{94}\text{Pu} + 2\beta^{-} + 2\bar{\nu}$$

We need to know something about the way in which neutrons interact with these nuclides, in order to understand how we can make as many as possible interact with the $^{235}_{92}$U nuclei. The neutrons produced in the fission of $^{235}_{92}$U have energies of about 1 MeV; they move through the uranium with speeds of the order of 10^{7} m s^{-1}. As such a fast-moving neutron approaches a uranium nucleus it is very unlikely to be captured by the nucleus. It is most likely to fly past rapidly, and may well leave the material altogether.

How can we make these neutrons interact more strongly with the uranium nuclei? The answer is to slow them down. If we can reduce their energies below the level of about 1eV, their interaction with $_{92}$U nuclei increases dramatically.

The probability that a nucleus will absorb a neutron is described by its **cross section**. We picture the absorbing nucleus as a circular disc which will absorb the neutron if it lies in its path. The area of cross section is measured in m² (SI units), or in the more traditional unit of the barn (1 barn = 10^{-28} m²).

The need for a moderator

In many nuclear reactors a material called the **moderator** is used to slow down the neutrons. This is a material with light nuclei which must not absorb neutrons. Typically, a **fast neutron** of energy 1 MeV collides perhaps 100 times with moderating nuclei; its energy decreases to about 1/40 eV as a result, and it is then known as a **thermal neutron**. (The significance of the figure of 1/40 eV is that this is the typical energy of particles at temperatures around room temperature.)

Materials used for moderators include carbon (in the form of graphite) and heavy water, D_2O, in which D (deuterium) is an isotope of hydrogen, $^{2}_{1}H$.

In the nuclear reactors known as **thermal** reactors (because they rely on a moderator to slow down the neutrons to thermal energies) the uranium fuel is mixed in some way with the moderating material. Usually, fuel rods are inserted into a large mass of moderator.

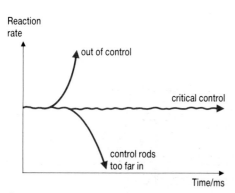

Fig 8.13 Critical control is tricky to maintain.

Control rods

The operation of a moderator gives us an idea of how further to control a nuclear chain reaction. The moderator slows neutrons down so that they interact more strongly with the uranium nuclei. If a chain reaction goes supercritical we need to be able to reduce the concentration of neutrons. We need to include some material which will absorb neutrons and so reduce the **neutron flux** within the uranium.

Cadmium is a suitable material. Cadmium nuclei have a large cross section for the capture of neutrons – about 10 times the cross-section of $^{235}_{92}U$ for thermal neutrons. Rods of cadmium are inserted into a nuclear reactor to slow or shut down the chain reaction; withdrawing the rods restarts or speeds up the reaction.

The delayed neutrons which arise from the decay of fission products have an important part to play in the control of practical chain reactions. Fig 8.13 shows how tricky it is in practice to control a chain reaction. With the control rods too far into the reactor, the reaction stops rapidly; with the rods too far out, the reaction rushes out of control. This can happen in a matter of milliseconds.

Fortunately it is possible to control the reaction by relying on the delayed neutrons. These are produced by reactions with half-lives of several seconds or even minutes. This is such a timescale that the control rods can more realistically be operated to control the reaction. The flux of delayed neutrons holds the balance between a runaway exponential reaction and a reaction which quickly grinds to a halt.

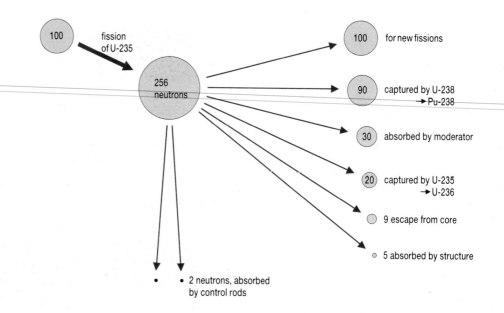

Fig 8.14 The life histories of neutrons in a reactor.

Fig 8.14 summarises the typical 'life histories' of neutrons within a reactor: 100 neutrons which cause fissions of $^{235}_{92}$U nuclei result in, on average, 256 further neutrons. The numbers on the figure indicate what happens to these neutrons. Note that, for a self-sustaining reaction, 100 must go on to initiate further fissions. Only about two of the 256 neutrons are absorbed by the control rods.

Fig 8.15 The structure of the core of a nuclear reactor.

In Fig 8.15, you can see how the fuel, moderator and control rods are arranged in a typical reactor core.

Can a power reactor become a bomb?

There are similarities between nuclear power reactors and nuclear bombs. There are also differences. For an effective bomb you need to ensure that all the fissile material undergoes fission in the minimum time possible. In practice the flux of neutrons increases by a factor of 10^{11} in about 1μs after the chain reaction reaches the critical state (see Fig 8.16). Most of the fissile nuclei are split in this time.

The figure also shows what would happen in a nuclear power reactor if the control rods were completely removed. In a power reactor the fissile material is spread out within the moderator; it is far less concentrated than in a bomb. Its temperature rises dramatically, but not to the extent of becoming like a bomb. The temperature levels off; the core of the reactor (fuel and moderator) may become molten. This is called a **meltdown.**

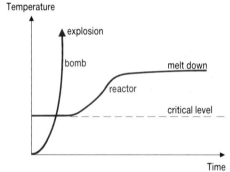

Fig 8.16 What happens when chain reactions go out of control.

QUESTIONS

Table 8.2

Nuclide	Cross section (barns)
$^1_1$H	0.332
$^2_1$H	0.000
^{53}Kr	0.009 2
B	759
C	0.003 4
O	0.000 270
Cd	2450
U	7.59

8.15 Calculate the kinetic energy E_k in MeV of a fast neutron moving at 10^7 m s^{-1}. (You may assume that the classical, non-relativistic equation for E_k applies.)

8.16 Calculate the speed of a thermal neutron of kinetic energy 1/40 eV.

8.17 When a nuclear reactor is operated continuously the control rods must be withdrawn gradually, over a period of weeks or months. What explanation can you give for this?

Table 8.2 shows figures for the cross sections of various nuclei for the capture of thermal neutrons. Use this data to answer the following questions.

8.18 Explain which materials would be useful for controlling a nuclear reactor.

8.19 Which of the elements listed would be useful for moderating fast neutrons?

8.20 Light water (H_2O) is unsuitable as a moderator, since ${}_1^1H$ nuclei absorb neutrons. Write an equation for this process. (A γ photon is produced together with the daughter nuclide.) How would light water compare with cadmium as a control material?

INVESTIGATION

You can use the *Nuclear Reactor Simulation* software (see Appendix A) to investigate the difficulty of operating a reactor. In particular, you can find out how the chain reaction is controlled using control rods.

If the reaction escalates to an unsafe level the automatic emergency shutdown procedure comes into operation, and you have failed the test!

8.4 POWER FROM FISSION

The contribution of nuclear power to world energy production increased dramatically in the 1980s. By 1986 it contributed about 20 per cent to electricity generation in industrialised countries – about 5 per cent of the world's primary energy consumption. Fig 8.17 shows the 20 countries with the highest nuclear shares of electricity generation.

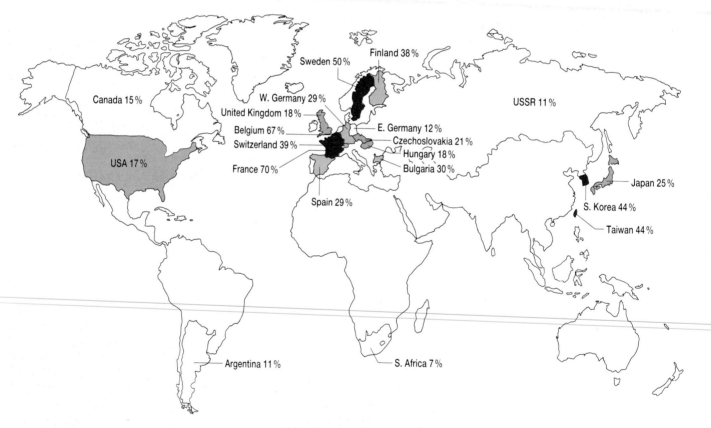

Fig 8.17 The contribution of nuclear power stations to electricity generation; the 20 biggest producers in 1986.

Energy from fission

Earlier, in question 8.6, you calculated an example of the energy released when a single ${}_{92}^{235}U$ nucleus splits. On average, each fission of a ${}_{92}^{235}U$ nucleus releases about 202 MeV. In what form does this energy appear? Fig 8.18 shows the approximate way in which the energy is shared out between different forms.

The **core** of a nuclear reactor (fuel, moderator, control rods and supporting structure) heats up as the energy released in fission is dissipated within it. The operating temperature of the core depends on the type and construction of the reactor.

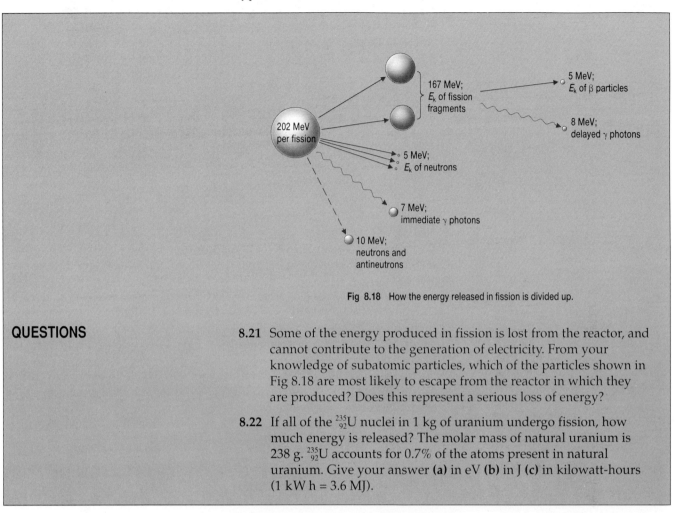

Fig 8.18 How the energy released in fission is divided up.

QUESTIONS

8.21 Some of the energy produced in fission is lost from the reactor, and cannot contribute to the generation of electricity. From your knowledge of subatomic particles, which of the particles shown in Fig 8.18 are most likely to escape from the reactor in which they are produced? Does this represent a serious loss of energy?

8.22 If all of the $^{235}_{92}$U nuclei in 1 kg of uranium undergo fission, how much energy is released? The molar mass of natural uranium is 238 g. $^{235}_{92}$U accounts for 0.7% of the atoms present in natural uranium. Give your answer **(a)** in eV **(b)** in J **(c)** in kilowatt-hours (1 kW h = 3.6 MJ).

The first occasion on which a controlled, self-sustaining chain reaction was established was on 2 December 1942. This first reactor was built in a converted squash court at the University of Chicago by a team led by

Fig 8.19 The world's first controlled, self-sustaining nuclear chain reaction; Chicago, 2 December 1942.

NUCLEAR FISSION REACTORS

Enrico Fermi. European attempts to do the same had been thwarted by the difficult working conditions during the Second World War. (You can see a full-scale replica of Fermi's reactor at the Science Museum in London.)

The Chicago reactor used 46 tons of uranium and uranium oxide fuel, graphite as moderator, and cadmium control rods. In Fig 8.19 a technician is withdrawing a control rod to start the reaction. The men on top of the reactor have bottles of cadmium solution; these would have been poured into the core to stop the reaction if it went out of control. At first it produced no more than 0.5 W of power, although later its power rose to 200 W.

QUESTION	**8.23** How many $^{235}_{92}U$ nuclei must undergo fission each second to generate a power of 0.5 W?

Coolant

The core of a reactor is hot; its temperature is perhaps 500°C, though this depends on the reactor type. Some reactors operate at temperatures as high as 1500°C.

The heat energy available in the core must be extracted if it is to be converted into electrical energy. A coolant flows around the core and carries heat energy to a steam generator. In Section 8.6 you will find out about the practical arrangements by which this is achieved in different reactors.

Coolant materials vary between reactors. They include water and carbon dioxide gas at high pressures. Reactors operating at high temperatures may use molten sodium as coolant.

The hot coolant passes into a steam generator, from where steam is passed to the turbines which generate electricity.

Fig 8.20 Hartlepool Power station.

8.24 Hartlepool Power Station (Fig 8.20) in North East England has two reactors, each of which generates 1500 MW of thermal (heat) power. This heat is transferred by the coolant (CO_2) to 16 steam generators, which feed two turbine generator units. The total electrical power generated is 1249 MW. Calculate the thermal efficiency of the power station – i.e. the efficiency with which thermal power is converted to electrical power.

8.5 NUCLEAR FUEL AND NUCLEAR WASTE

The story of nuclear power begins in uranium mines; it ends with the disposal of radioactive waste. In this section you can find out a little of the processes involved in handling these radioactive materials.

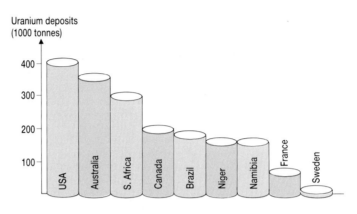

Fig 8.21 The world's major deposits of uranium.

Extracting and enriching uranium

Uranium ore is mined principally in the USA, the USSR, Canada, Namibia and Niger. Australia also has large reserves (see Fig 8.21). The ore may have as little as 0.4 per cent of uranium when it is extracted by surface or

Fig 8.22 A uranium mine in Queensland, Australia.

underground mining (see Fig 8.22). The process of producing uranium oxide in the form known as **yellow cake** is summarised in Fig 8.23.

Fig 8.23 Extracting uranium oxide from ore.

Natural uranium contains only 0.7 per cent of the useful fissile isotope $^{235}_{92}U$. Most of the remainder is $^{238}_{92}U$. Some reactors use **enriched uranium** as fuel; this has a higher proportion, about 2.3 per cent, of $^{235}_{92}U$. In the process of enrichment the uranium is converted to the volatile gas UF_6, known as **hex**. Enrichment can only be achieved by physical means, since all isotopes of uranium are chemically virtually identical.

The most common technique for the separation of isotopes depends on the faster diffusion through a membrane of the lighter $^{235}_{92}U$ isotope. An alternative method involves the centrifuging of hex gas; the heavier isotope drifts to the outer perimeter of the centrifuge. A third technique involves ionising $^{235}_{92}U$ hex molecules using a tuned laser. The details of enrichment methods are kept secret because of their military importance.

The process of uranium enrichment is technologically very difficult since hex is a highly corrosive, reactive gas. It requires very careful handling.

Fuel fabrication

Fig 8.24 Uranium fuel **(a)** pellets of UO_2

Reactors may use uranium as fuel in the form of uranium metal or uranium dioxide. UO_2 powder is produced from the enriched hex and then pressed into the form of pellets. Some are shown in Fig 8.24(a). Fig 8.24(b) shows fuel rods for a reactor; these are stainless steel tubes containing 60 to 70 pellets of UO_2.

Fuel reprocessing

When, after perhaps four years, the amount of $^{235}_{92}U$ remaining in the fuel rods drops below the level at which a chain reaction can be maintained, the fuel must be reprocessed. It contains some valuable materials – $^{235}_{92}U$, $^{238}_{92}U$ and $^{239}_{94}Pu$. However, it also contains the fragments which result from fission, many of which are highly radioactive.

The chemical process whereby the useful materials are extracted is illustrated in Fig 8.25. It is essential to recover the plutonium, for not only is it a useful bomb-making material but it is also exceedingly toxic and should not be released into the environment. There is also always a danger

(b) Magnox fuel rods.

NUCLEAR FISSION REACTORS

when handling plutonium that a critical concentration might be achieved, with a catastrophic explosion resulting.

Fig 8.25 The principles of nuclear fuel reprocessing.

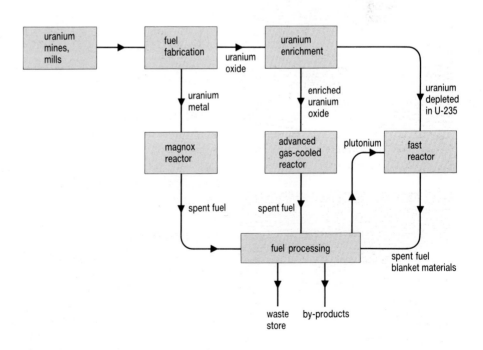

Fig 8.26 The nuclear fuel cycle.

There are many problems associated with the production of nuclear fuel from uranium ore. At all stages, dangerous waste materials are produced. Some of these problems are discussed briefly above.

Read about the nuclear fuel cycle (Fig 8.26) in some of the literature listed in Appendix A. List some of the waste materials produced by the nuclear industry. What are the hazards associated with the different wastes?

At the end of this section you can read more about the categorisation and handling of nuclear waste.

Plutonium as a fuel

$^{239}_{94}Pu$ is a transuranic element produced in uranium fission reactors; the equations for this are given at the beginning of this chapter. $^{239}_{94}Pu$ is also used as a fuel in some fission reactors and in some bombs.

Plutonium acts as a nuclear fuel in a very similar way to uranium. $^{239}_{94}Pu$ nuclei split when they absorb a single neutron. However, there is a difference between plutonium fission and uranium fission; plutonium requires fast neutrons and cannot be split by thermal neutrons. This means that the technology of fast reactors, which use plutonium, is rather different from that of uranium-fuelled reactors.

To understand why $^{239}_{94}Pu$ will not split with slow-moving thermal neutrons we must look more closely at the process of fission. The first step in the process is the capture of a neutron:

$$^{239}_{94}Pu + {}^{1}_{0}n \rightarrow {}^{240}_{94}Pu$$

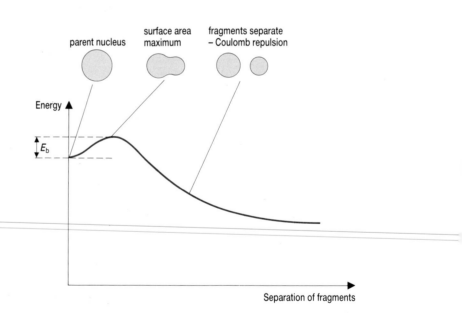

Fig 8.27 Energy changes as a nucleus splits.

This results in an energetic nucleus, $^{240}_{94}Pu$, which distorts and may split. However, in order to split, the nucleus must have enough energy to create extra surface area. Fig 8.27 shows how the energy of the nucleus changes as it splits. The nucleus must have enough energy to overcome the barrier energy E_b. In the case of the fission of $^{235}_{92}U$ the nucleus formed after neutron capture, $^{240}_{92}U$, has sufficient energy to split even for a thermal neutron. $^{240}_{94}Pu$, however, does not have enough energy to overcome the barrier unless it captures a fast neutron with energy at least 1 MeV.

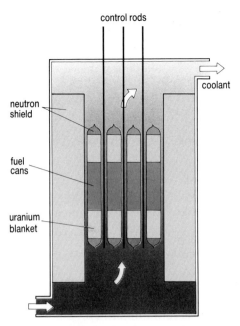

control rods

coolant

neutron shield

fuel cans

uranium blanket

Fig 8.28 The structure of the core of a fast reactor.

Breeding plutonium

Plutonium is not an element found in nature. All the plutonium which exists has been made in reactors – some for military use, and some as a by-product of the fission of uranium in civil reactors.

Fast reactors have been designed to use this plutonium as fuel – you may have heard them described as breeder reactors. This is because they are designed so that, as part of their operation, they produce more plutonium than they use. This may seem like something for nothing; how can a power station produce more fuel than it uses?

$^{239}_{94}\text{Pu}$ is formed when a nucleus of $^{238}_{92}\text{U}$ captures a neutron. The core of a fast reactor contains plutonium (see Fig 8.28) which becomes hot through fission involving fast neutrons. Surrounding the core is a blanket of uranium, which mops up some of the neutrons which escape from the plutonium. $^{238}_{92}\text{U}$ is converted to $^{239}_{94}\text{Pu}$. Since each plutonium fission event produces on average 2.91 neutrons, and only one is required to perpetuate the chain reaction, there are plenty of spare neutrons for breeding more plutonium.

Fast reactors make use of $^{238}_{92}\text{U}$, which makes up 99.3 per cent of natural uranium. This is a much more effective use of uranium than thermal reactors, which only use the 0.7 per cent which is $^{235}_{92}\text{U}$.

Prototype fast reactors have operated for years; the British reactor at Dounreay in the north of Scotland can provide 250 MW of power. However, problems with the technology of these advanced reactors, concerns about their safety and an abundant world supply of uranium for thermal reactors have combined to limit the use made of fast reactors.

QUESTIONS

Table 8.3

Nuclide	Mass
$^{239}_{94}\text{Pu}$	239.052 17
$^{145}_{56}\text{Ba}$	144.926 94
$^{93}_{36}\text{Kr}$	92.931 12
$^{1}_{0}\text{n}$	1.008 665

8.25 Fast reactors use plutonium $^{239}_{94}\text{Pu}$ as fuel. Explain how you would expect the construction of a fast reactor to differ from a thermal reactor. Which material would not be necessary?

8.26 When $^{239}_{94}\text{Pu}$ undergoes fission the average number of neutrons released per fission event is 2.91. For $^{235}_{92}\text{U}$ the corresponding figure is 2.47. Explain why this would help in the design of a fast reactor.

8.27 Calculate the energy released when a single nucleus of $^{239}_{94}\text{Pu}$ captures a neutron and splits, if the fission products are $^{145}_{56}\text{Ba}$ and $^{93}_{36}\text{Kr}$. Mass values are given in Table 8.3.

Handling nuclear waste

From the previous discussion you will be aware that the nuclear industry produces a large amount of radioactive waste materials. Considerable effort has been put into solving the problem of safe disposal of this waste. Difficulties still remain.

In deciding how to dispose of waste you need to consider several factors: the activity of the waste, its half-life and its chemical properties. Plutonium $^{239}_{94}\text{Pu}$ is an α emitter, and it is relatively simple to provide protection against α radiation. However, this isotope's half-life is 24 000 years and it is extremely corrosive and toxic; any containment system for plutonium must be secure for many thousands of years.

Table 8.4 lists the categories into which nuclear wastes are divided, together with ways in which they may be disposed of. The question of dumping waste is an on-going problem for the industry, which arouses considerable public concern.

Table 8.4　Disposal of nuclear wastes.

Waste category	Examples	Methods of disposal
low-level	discarded protective clothing, used wrapping materials	liquids and gases released to environment, solids buried on land or at sea
intermediate-level	irradiated fuel cladding, reactor components, chemical residues	concrete stores, deep trenches
high-level	fission fragments from reprocessed nuclear fuel	liquid stored in steel-lined, water-cooled tanks

ASSIGNMENT

Read about the management of radioactive waste in some of the literature listed in Appendix A and answer the questions which follow.

8.28　Methods of disposal may be classified as **containment** or **dispersal**. Explain these terms. Which of the methods of disposal listed in Table 8.4 involve dispersal into the environment?

8.29　Low-level waste is defined as having activity below 4×10^9 Bq for α emitting waste, and below 1.2×10^{10} Bq for β and γ emitting waste. Explain why the limit is set at a lower level for α active material.

8.30　Suppose that a low-level nuclear waste dump was established in a field behind your home. What would you expect to see happening there? What fears might you and your neighbours have about the dump? Do you think such fears are enough to justify outright opposition to such a dump?

8.6　PRACTICAL NUCLEAR REACTORS

You should now be familiar with the principles of operation of nuclear reactors and some of the problems of providing them with fuel, controlling their safe operation and disposing of dangerous waste materials. Now you are in a position to understand the similarities and differences between the different types of fission reactors in use at present. At the end of this section you can read about a fission chain reaction which occurred in natural uranium deposits in Africa, more than a billion years ago.

ASSIGNMENT

Figs 8.29 to 8.34 show the construction of various different nuclear power reactors. Examine these diagrams; look for the following: fuel elements, moderator, control rods, steam generator, pressure vessel, turbine generator (not all reactors have all of these). Trace the path of circulation of the coolant.

From these diagrams, and by studying the literature produced by the nuclear industry and other sources, you can find out further details of these reactors. Draw up a table to show, for each reactor, the following points:

- Is it a **thermal** or **fast** reactor?
- What **fuel** is used?
- What materials are used for the **moderator**, **control rods** and **coolant**?
- Which **country** developed it?

Fig 8.29 A MAGNOX reactor; the fuel is clad in a magnesium alloy called magnox.

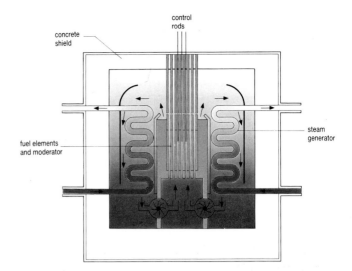

Fig 8.30 An advanced gas-cooled reactor (AGR).

Fig 8.31 A pressurised water reactor (PWR). The Sizewell B power station will be a PWR.

Fig 8.32 A CANDU reactor, developed in Canada.

Fig 8.33 An RBMK reactor, the kind which exploded at Chernobyl on 26 April 1986.

Fig 8.34 A fast reactor.

NUCLEAR FISSION REACTORS

8.31 Which types of reactors currently make major contributions of electricity to the UK National Grid? Which are likely to contribute in the future?

8.32 What are the functions of the pressure vessel and the concrete shield?

8.33 You may have heard of nuclear-powered ships and submarines. Which type of reactor is used in these vessels?

ASSIGNMENT

A natural fission reactor

For thirty years it was assumed that the first nuclear chain reaction to occur on the Earth was that set up by Fermi in Chicago in 1942. However, it has now been established that a natural reactor had existed in natural uranium deposits 1.8 billion years ago.

Evidence for this came in an interesting way. Natural uranium from Gabon was exported to France; an examination of the isotopic content showed that the proportion of uranium-235 was slightly lower than normally found (0.7171 per cent, compared to 0.7202 per cent). This small difference was enough to arouse the interest of some of the scientists working on fuel processing, and they looked more closely into the composition of the ore. They found traces of the fission products of uranium, in higher proportions than in normal uranium ore. Their suspicion was that, at some time in the geological history of the uranium, some of it had undergone a fission reaction. But how could a chain reaction have been established in natural uranium?

The seam of ore which was being extracted was unusually rich in uranium – up to 10 per cent. Geological conditions had conspired to accumulate large quantities in a small area. In addition, the proportion of uranium-235 would have been higher at an earlier date. The water of crystallisation of the minerals in the ore might have acted as a moderator.

It is now believed that a natural fission chain reaction must have taken place in the ore, at a time approximately 1800 million years ago. It may have run for 10^5 to 10^6 years, emitting a thermal power of tens of kilowatts. (Any greater power would have led to the evaporation of the water required as a moderator.) In the course of its lifetime it would have consumed a similar amount of uranium as a present day power reactor consumes in a year.

8.34 The uranium ore was found to be depleted in U-235. What fraction had been removed by fission?

8.35 Explain why the proportion of U-235 was higher in the past. Estimate the fraction of uranium which was U-235 at the time of the chain reaction. (Half-lives: for U-235 it is 710 million years; for U-238 it is 4500 million years.)

8.36 Estimate the mass of uranium consumed in the lifetime of the chain reaction.

(You can read more about this natural chain reaction, and the way in which its history was unravelled, in the July 1976 issue of *Scientific American*.)

SUMMARY

Some massive nuclei split into two large fragments, either spontaneously or after capturing a neutron. Since two or more neutrons are released in fission it is possible to establish a chain reaction. A minimum amount of material, the critical mass, is needed for this to happen. Energy is released in the process.

In a fission bomb the chain reaction escalates rapidly and the energy is released in a fraction of a second. In a power reactor the rate of reaction is controlled by neutron-absorbing control rods. In thermal reactors neutrons are slowed down to thermal energies by the moderator, which absorbs energy from the neutrons without absorbing the neutrons themselves. In fast reactors there is the possibility of breeding plutonium from uranium.

The processing of nuclear fuel and waste requires advanced techniques for handling these dangerous materials.

Chapter 9

NUCLEAR FUSION

In the last chapter you learnt how large nuclei can split and thereby release energy. It turns out that small nuclei can combine, or fuse together, to give larger nuclei. This process of nuclear fusion also results in the release of energy. In this chapter you can find out about nuclear fusion, its occurrence in nature and the prospects of harnessing it as an almost limitless source of energy.

LEARNING OBJECTIVES

After studying this chapter you should be able to:

1. calculate the energy released when two light nuclei fuse;

2. explain why nuclear fusion reactions can occur at high temperatures;

3. outline the practical problems involved in developing fusion power sources.

9.1 THE POSSIBILITY OF FUSION

By now you should be familiar with the idea that nuclear reactions (radioactive decay, fission, fusion) can occur if the final particles have less mass than the parent particles. The difference in mass appears as energy; the equation $E = mc^2$ shows how to convert from mass to energy. The product particles are more tightly bound than the initial particles; they have greater binding energy.

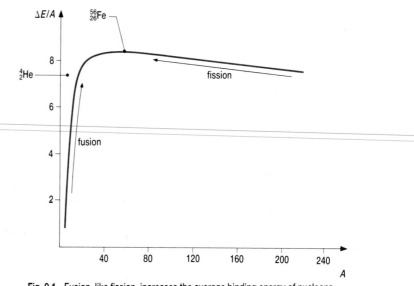

Fig 9.1 Fusion, like fission, increases the average binding energy of nucleons.

Fig 9.1 reproduces the binding energy curve of Fig 4.5. Recall that the most tightly bound nuclei are in the region of the iron nuclide $^{56}_{26}$Fe. The figure also shows the way in which massive nuclei such as $^{235}_{92}$U can become more stable by the process of fission. An arrow shows how light nuclei can

BEFORE   AFTER

Fig 9.2 Two deuterium nuclei fuse to give a helium nucleus.

become more stable by increasing their nucleon number A. They can do this by fusing together.

Energy release

We can calculate the energy released in a fusion reaction in an identical way to previous calculations for other nuclear processes. Simply compare masses before and after the reaction, and use $E = mc^2$. Here is an example of an equation for a fusion reaction, as shown in Fig 9.2:

$$^{2}_{1}H + {}^{2}_{1}H \rightarrow {}^{4}_{2}He + Q$$

$^{2}_{1}H$ represents a deuterium nucleus, an isotope of hydrogen. Two deuterium nuclei can fuse to give a helium nucleus, $^{4}_{2}He$. Q represents the energy released in the reaction.

Table 9.1 gives the masses of several light nuclei, including those of interest here. Rewriting the equation in terms of masses gives:

$$2.014\ 102 + 2.014\ 102 \rightarrow 4.002\ 604 + Q$$

or $\qquad 4.028\ 204 \rightarrow 4.002\ 604 + Q$

Since the mass is less after the reaction than before, we have established that this reaction can indeed occur. The mass decrease is:

$$4.028\ 204 - 4.002\ 604 = 0.0256\ u$$

which corresponds to 23.8 MeV of energy. The binding energy of each nucleon has increased by about 6 MeV. (You will recall, if you look back to Fig 4.5, that $^{4}_{2}He$ is an unusually tightly bound nuclide.)

Table 9.1

Nuclide	Mass/u
$^{0}_{-1}e$	0.000 549
$^{1}_{1}H$	1.007 825
$^{1}_{0}n$	1.008 665
$^{2}_{1}H$	2.014 102
$^{3}_{1}H$	3.016 049
$^{3}_{2}He$	3.016 030
$^{4}_{2}He$	4.002 604
$^{7}_{3}Li$	7.016 005
$^{7}_{4}Be$	7.016 931
$^{8}_{5}B$	8.024 612

QUESTIONS

9.1 Which of the following fusion reactions can occur? In those which are possible, how much energy is released?

(a) $^{2}_{1}H + {}^{1}_{1}H \rightarrow {}^{3}_{2}He + Q$

(b) $^{1}_{1}H + {}^{7}_{3}Li \rightarrow 2({}^{4}_{2}He) + Q$

(c) $^{1}_{1}H + {}^{7}_{4}Be \rightarrow {}^{8}_{5}B + Q$

9.2 In one of the fusion reactions which occur in stars, protons fuse to give deuterium. The equation given below is incomplete; which two particles are required to complete it? (Hint: remember, you must make both sides balance in terms of mass number, charge and the numbers of particles.)

$$^{1}_{1}H + {}^{1}_{1}H \rightarrow {}^{2}_{1}H + ? + ? + Q$$

9.3 In the reaction $^{2}_{1}H + {}^{1}_{0}n \rightarrow {}^{3}_{1}H$, a deuterium nucleus captures a neutron and becomes a tritium nucleus. (Tritium is another isotope of hydrogen.) Calculate the energy released.

Tritium is radioactive. Would you describe it as proton-rich or neutron-rich? What kind of decay would you expect it to undergo? Write down an equation for the decay of tritium and calculate the energy released in this decay.

9.4 Another reaction which produces tritium is $^{2}_{1}H + {}^{2}_{1}H \rightarrow {}^{3}_{1}H + {}^{1}_{1}H$. The energy produced in this reaction is 4 MeV. If this energy appears as kinetic energy of the two product particles, how would you expect it to be shared between them?

Fusion in the stars

It is fortunate for us that fusion is possible. The energy we receive from the Sun is a result of fusion, and the elements which form the basis of our material world were formed by fusion in stars.

The Sun consists largely of hydrogen and helium. Within the Sun, where the temperature is millions of kelvins, there is a constant fusion of nuclei. One of the main reaction cycles is the **hydrogen cycle**, or **proton–proton cycle**. These are the reactions involved:

$$^1_1H + ^1_1H \rightarrow ^2_1H + \beta^+ + \nu$$
$$^2_1H + ^1_1H \rightarrow ^3_2He$$
$$^3_2He + ^3_2He \rightarrow ^4_2He + 2(^1_1H)$$

In each of these reactions energy is released. The overall effect is that four protons have combined to give a helium nucleus ^4_2He (and two positrons and two neutrinos.) The reactions of this cycle give us most of the energy which we receive from the Sun.

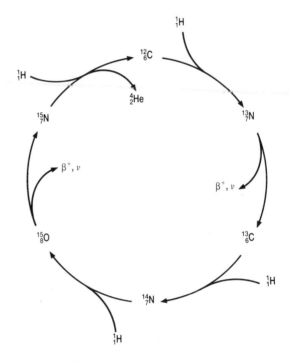

Fig 9.3 The carbon cycle, a fusion process found in stars.

There is a second important cycle of fusion reactions which occurs in the Sun, the **carbon cycle**. This is shown in Fig 9.3. Again, the net result is that four protons give rise to a single helium nucleus.

9.2 THE PROCESS OF FUSION

There is a problem with the discussion so far. We have seen that nuclei fuse to increase their binding energy, and the greatest binding energy per nucleon is for nuclei around $^{56}_{26}Fe$. So why don't all nuclei fuse together to form iron? Why is the Sun made mostly of hydrogen and helium, rather than being a lump of iron? To understand the answer to this we must think about the details of the process of fusion.

NUCLEAR FUSION

Sticking nucleons together

The process of fusion is rather like fission, seen in reverse. Two small particles come together to form a larger one (Fig 9.4). From what you already know about the forces between nucleons, you can work out the conditions necessary for fusion. Then we can go on to discuss how it might be possible to use nuclear fusion as a practical energy source.

Think about the process shown in Fig 9.4. In (a) the two protons are far apart, with little force acting between them. In (b) they are closer together and repel one another. In (c) they have formed a nucleus, and no longer repel each other.

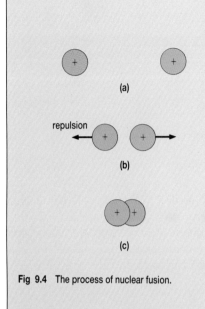

Fig 9.4 The process of nuclear fusion.

9.5 What force causes the two protons to repel one another? If their separation is 10^{-14} m, calculate **(a)** the force of repulsion, and **(b)** the potential energy in eV of one proton in the field of the other.

To bring two protons together in this way they must be given a lot of kinetic energy. You must raise the temperature of the protons until they are moving so fast that, if they meet, they approach close enough for the attractive nuclear force to overcome their electrostatic repulsion. But what temperature is required?

From the kinetic theory of gases, you may know that, at temperature T, the kinetic energy of a simple particle like a proton is given by:

$$E_k = \tfrac{3}{2} kT$$

where $k = 1.38 \times 10^{-23}$ J K^{-1} is Boltzmann's constant.

9.6 As one proton approaches the other it slows down. Its kinetic energy is converted to potential energy. For the protons to fuse, their kinetic energy must be at least as great as the potential energy you calculated above. Use the equation for E_k to calculate the temperature necessary for fusion.

9.7 Explain why a higher temperature would be needed for larger nuclei such as $^{16}_{8}$O to fuse.

The fourth state of matter

How can we achieve such high temperatures? Nuclear fusion requires temperatures in excess of 100 000 000 K. What is matter like at such a temperature? All matter is gaseous above about 6000 K. What happens if we make it even hotter?

The answer is that a **plasma** is formed. In a sufficiently hot gas, the atomic electrons break free from the hold of the nucleus and the gas becomes a fluid mixture of electrons, positively charged ions and even free nuclei. (A fluorescent light tube contains a low-temperature plasma.)

For technologists, the problem with plasma is how to contain it. Stars are made of plasma; containing plasma is like bottling a star. This is the subject of Section 9.3. Before that we will look at fusion in stars.

The life history of the Universe

The story of the Universe is a story of fusion. From the initial Big Bang, 15 000 million years ago, hydrogen and helium formed. Their thermal motion was too great for them to fuse further to give more massive nuclei. As the resulting gas cooled stars were formed (Fig 9.5). Within stars, more helium is formed by the proton–proton cycle (see Section 9.1). If the temperature is high enough, the carbon cycle may also operate.

The final phases of the evolution of a star depend on its mass. Less massive stars (like the Sun) expand to become **red giants,** with a hot, dense core. Within a red giant, fusion reactions occur such as $3(^4_2\text{He}) \rightarrow {}^{12}_{6}\text{C}$ and $4(^4_2\text{He}) \rightarrow {}^{16}_{8}\text{O}$. This carbon and oxygen blows out into space as part of the stellar wind.

Fig 9.5 The evolutionary history (from left to right) of stars.

Heavier stars blow up as supernovae. The energy released by fusion is enough to raise the temperature to levels needed to produce elements such as iron, silicon and neon, and even uranium. These are spread out into space in the ensuing explosion, and then become part of stars and planetary systems which form later. Our understanding of fusion has helped us to understand the life history of the stars and the origins of the elements from which the Earth is made.

QUESTION

9.8 (a) The Sun has a mass of 2×10^{30} kg. If its original mass consisted of 10% protons and 90% helium, how many protons did it contain? ($m_p = 1.67 \times 10^{-27}$ kg.)

(b) Fusion reactions produce 6.4 MeV per proton. If the Sun radiates energy at a rate of 4×10^{26} W, estimate the lifetime of the Sun. (1 year is approximately 3×10^7 s.)

(c) If the Sun is 5000 million years old, how much longer will it last?

9.3 FUSION AS AN ENERGY SOURCE

Fusion has already been used – in some nuclear weapons. This is a very crude, uncontrolled use of a fusion reaction. The possibility of a controlled fusion reaction as a practical energy source is still a distant prospect, although some of the technical problems have now been solved.

The hydrogen bomb

Hydrogen fusion bombs use a mixture of deuterium and tritium, which are forced together at high temperatures achieved by first detonating a fission bomb (Fig 9.6). In a fission–fusion–fission bomb there is an outer blanket of natural uranium. This captures the neutrons released by the fusion reaction, and more fission results; there is a greater radioactive fallout, and the bomb is considerably more lethal.

Neutron bombs are battlefield nuclear weapons – hydrogen fusion bombs designed to maximise the production of neutrons. The blast and heat generated are much less than for an ordinary H-bomb. The bomb is detonated at a few hundred metres above ground; people in the vicinity are killed by the intense neutron radiation, but buildings and vehicles are relatively unscathed.

Fig 9.6 The construction of a fusion bomb.

The potential of fusion power

Deuterium is an abundant resource. It represents one in 6700 hydrogen atoms; one cubic metre of water contains 10^{25} deuterium atoms. Tritium, which is produced from lithium, is also abundant, although not as much as deuterium. It is estimated that, with the development of appropriate techniques, fusion power could satisfy energy needs for at least 200 times the lifespan of the thermal nuclear reactor programme.

The problems involved are immense. It is necessary both to contain a suitable plasma and to extract energy from it.

Plasma containment

A plasma may be formed at a temperature of 10^8 K. It cannot simply be placed in a container, since any contact with solid walls would cool it instantly. So how might it be confined?

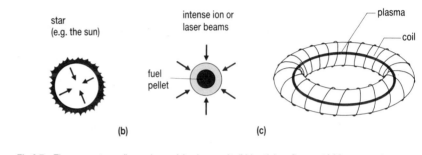

(a) **(b)** **(c)**

Fig 9.7 Three ways to confine a plasma **(a)** using gravity **(b)** inertial confinement **(c)** in a magnet 'doughnut'.

Fig 9.7 illustrates three ways in which a plasma may be contained. Nature's way, a star, relies on the gravitational attraction to hold the

Fig 9.8 The Joint European Torus at Culham in Oxfordshire is a large scale experiment to confine plasma magnetically. (A torus is a doughnut or wedding-ring shape.)

Fig 9.9 Inside the JET vacuum chamber.

plasma together. This requires a vastly greater mass of plasma than we could deal with on the Earth.

Inertial confinement relies on intense beams of laser light or ions to compress a fuel pellet whilst heating it. This is an approach currently receiving considerable attention.

Magnetic confinement, or 'bottling', uses the fact that the particles of a plasma are charged. They move in helical paths around the magnetic field lines. With a cleverly designed magnetic field, the plasma will circulate endlessly, held away from the walls of the containing vessel. The Russian Tokamak system (designed by Andrei Sakharov and Igor Tamm) and the European JET system (Figs 9.8 and 9.9) are examples of experimental systems which have achieved temperatures in excess of 10^7 K for fractions of a second.

Extracting energy

Techniques for extracting energy from fusion reactors are also being developed. One idea, shown in Fig 9.10, is to surround a reactor with a blanket of lithium. This would trap neutrons produced in the fusion reaction, and take up their energy. Some of the lithium would be converted to tritium, which could then be extracted and used as fuel.

The hot (molten) lithium would heat water to give steam, and then the process of electricity generation is conventional.

Fig 9.10 The principles of construction of a fusion reactor.

QUESTION	9.9 If higher temperatures can be achieved, it might be possible to run a fusion reactor which uses only deuterium as fuel, eliminating the need for tritium. This would vastly extend our fuel reserves – perhaps giving us enough energy to last 20 000 million years! Explain why the fusion reaction $^2_1\text{H} + {}^2_1\text{H}$ requires a higher temperature than $^2_1\text{H} + {}^3_1\text{H}$.

SUMMARY

Light nuclei can increase their binding energy, and hence their stability, by fusing together. To do this they must overcome the Coulomb repulsion barrier between them. High temperatures ($\sim 10^8$ K) are required to give them sufficient kinetic energy.

For fusion to become a useful power source, it is necessary to develop techniques for containing high-temperature plasma. Stars are plasma held together by gravity. Experiments are proceeding to try out methods of inertial confinement using intense laser beams, and magnetic bottling. There are also problems to be solved concerning the extraction of heat from a fusion reactor.

Chapter 10

NUCLEAR PHYSICS TODAY

In this chapter you are asked to use your knowledge of nuclear physics to look at some of the present (and future) problems of physics.

10.1 MONITORING CHERNOBYL

Fig 10.1 Monitoring environmental contamination by radioactive waste on a Cumbrian beach near Sellafield.

The nuclear industry produces large amounts of radioactive waste materials. (You looked at the nature of these in Chapter 8.) Usually these are disposed of in ways which are considered to be safe. It is important that efforts are put into monitoring the effects these wastes have on the environment (see Fig 10.1). In discharging waste into the sea, for example, it is hoped that the waste becomes gradually more and more dispersed.

Not everyone thinks that this is a fit way to treat our environment. It may be that, for example, hazardous materials accumulate and become locally concentrated by some unknown mechanism. This uncertainty has led to many protests against the dumping activities of the nuclear industry; some protests have been of a very active nature (Fig 10.2).

These waste discharges from the nuclear power industry are deliberate, and controlled. Some other releases from nuclear power plants have not been deliberate. In the 1950s there was a large release of radioactive material from Windscale (Sellafield) in Cumbria. The United States nuclear power programme was largely halted after the accident at Three Mile Island. Then there was the accident at Chernobyl.

At 1.25 on the morning of April 26th 1986 the Number 4 reactor at Chernobyl, in the Ukraine, blew up. Operators were carrying out tests to see if it was possible to use the kinetic energy of a running down turbine to generate power for auxiliary plant. In the course of the test six different

Fig 10.2 Protestors attempt to block pipes carrying radioactive waste into the Irish Sea.

Fig 10.3 The devastated Number 4 reactor at Chernobyl.

Fig 10.4 Checking vehicles for radioactive contamination in the aftermath of the Chernobyl disaster.

operating rules were violated; by the time the operators decided to drop in the control rods to stop the fission reaction, the rods reactor core was so distorted that the rods would not go down to the bottom, and the reactor blew up. Early pictures showed the devastation (Fig 10.3). The Number 4 reactor was devastated and has since been completely entombed in a concrete shield more than one metre thick. Dozens of people have died as a direct result. These included workers at the plant, firemen who fought the ensuing blaze and people who lived in the neighbourhood. In the long term a cautious estimate is that the radiation released will result in perhaps 4000 additional deaths amongst members of the public. Extensive precautions were taken in the vicinity of the power plant to decontaminate vehicles and keep down the dust (Fig 10.4).

The United Kingdom experienced some consequences of the Chernobyl accident. A radioactive plume of dust blew westwards across Europe and rainfall brought the dust down in some parts of the country. For many hill farmers, whose sheep became contaminated, the effects were devastating. The map of Fig 10.5 shows the route of the dust cloud during the days following the accident. (Extensive international monitoring was able to show that the cloud drifted all the way round the world, and was still detectable as it drifted across Europe for a second time, several weeks later.)

Fig 10.5 The course of the radioactive dust cloud from Chernobyl as it drifted across Europe in the days following the explosion of 26 April 1986.

INVESTIGATION

At Ackworth School near Pontefract, West Yorkshire, pupils monitored the passage of the Chernobyl radiation (Fig 10.6). They were able to detect a significant rise in radiation levels above the usual background level as the cloud passed northwards over England.

Few people were sufficiently prepared to record radiation levels at the time, although it was perfectly possible to do so using school equipment. In this assignment you are asked to devise and test your own system for monitoring such an increase in radiation levels. Hopefully you will never have to use such a system for real. (You can read about another school's methods in 'A school investigation into Chernobyl fallout' by R D Plant in *Physics Education*, vol. 23 p. 26, January 1988.)

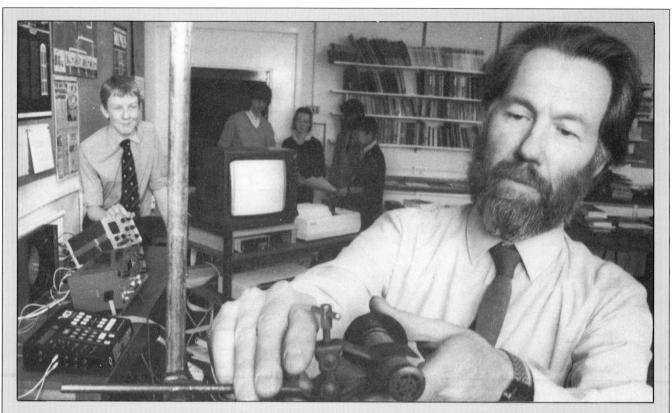

Fig 10.6   At Ackworth School, the passage of the radioactive dust cloud from Chernobyl was monitored.

Think about the measurements you will have to make and the time intervals over which you will have to record data. To give a realistic picture of the radiation you might be monitoring, suppose that the radiation level rises over a period of 12 hours to a level three times the background, and then falls back again over a similar time interval.

QUESTIONS

10.1 What time intervals will you record over, and how many counts will you expect?

Think about how you are going to record this information. The pupils at Ackworth school used a Geiger counter in conjunction with a VELA data-logger, which recorded their data. Do you want to have to record data manually over a period of 24 hours? Can you persuade your partner to do so? Can you use a data-logger or computer to do this for you?

10.2 How will you detect the radiation and how will you record the data? Draw up an experimental design and set it up in the laboratory to detect background radiation.

In the absence of a convenient nuclear accident, it is necessary for you to test your system. It is not essential that your test lasts 24 hours but you should ensure that you can detect and record significant variations in radiation levels over a suitable time interval.

10.3 How will you test your system? Carry out tests and write a report to show how your system can be used in the event of any future nuclear accident.

10.4 How could your experimental arrangement be adapted to measure variations in background radiation levels across the United Kingdom?

10.2 PULSARS, NEUTRON STARS, BLACK HOLES AND LITTLE GREEN MEN

In Chapter 2 you looked at the forces which hold the nucleus together. The strong nuclear force and the electrostatic Coulomb force are both important. One of the conclusions we came to was that gravity is not significant on the nuclear scale.

However, if we turn our attention to the astronomical scale we find that the gravitational attraction between nucleons can be very important indeed. We can see just how important this is by considering the semi-empirical binding energy formula (Chapter 4).

For a very massive nucleus, we can calculate its gravitational potential energy as follows. Consider a nucleus of mass M and radius r. Its gravitational energy E_{grav} is given by:

$$E_{\text{grav}} = \frac{3}{5} \cdot \frac{GM^2}{r} \tag{1}$$

(Here, G is the universal gravitational constant.) If the nucleus is made up of A nucleons, each of mass m and radius r_0, we can substitute $M = Am$ and $r = r_0 A^{1/3}$. This gives:

$$E_{\text{grav}} = \frac{3}{5} \cdot \frac{Gm^2A^2}{r_0 A^{1/3}} \tag{2}$$

The gravitational binding energy per nucleon is thus:

$$\frac{E_{\text{grav}}}{A} = \frac{3}{5} \cdot \frac{Gm^2 A^{2/3}}{r_0} \tag{3}$$

It turns out that, for massive nuclei made only of neutrons, the gravitational attraction is enough to hold them together. You can estimate the size of such a pure-neutron nucleus by adding gravitational energy to the semi-empirical binding energy formula.

For this nucleus, we can ignore the surface and Coulomb energy terms. This leaves us with an equation consisting of volume, symmetry and gravitational terms:

$$\text{binding energy per nucleon} = a - \frac{d(N-Z)^2}{A^2} + \frac{3}{5} \cdot \frac{Gm^2 A^{2/3}}{r_0} \tag{4}$$

Since, in this case, $A = N$, the equation becomes:

$$\text{binding energy per nucleon} = a - d + \frac{3}{5} \cdot \frac{Gm^2 A^{2/3}}{r_0} \tag{5}$$

Notice that the symmetry term is negative; this is what tends to make such a 'nucleus' unstable. To be stable, it must have positive binding energy; that is, we must have:

$$\frac{3}{5} \cdot \frac{Gm^2 A^{2/3}}{r_0} > d - a \tag{6}$$

From the values of the coefficients in Table 4.1 (page 48) you can see that $d - a = 5.3$ MeV. The inequality of equation (6) can be solved to show that A must be greater than about 10^{55}. The nucleus we have been talking about contains 10^{55} neutrons; it is an astronomically large nucleus, a neutron star.

The discovery of neutron stars

In the passage which follows you can read about the evidence for the existence of neutron stars, and how these strange astronomical nuclei are related to pulsars and black holes. The passage is an extract from *The Quantum Universe* by Tony Hey and Patrick Walters, Cambridge University Press, 1987.

Fig 10.7 Jocelyn Bell Burnell, the astronomer who first detected a pulsar.

The observational evidence for neutron stars is associated with the discovery of 'pulsars' by Jocelyn Bell (Fig 10.7), a research student of Anthony Hewish in Cambridge, UK, in 1967. Pulsars are rapid and remarkably regular radio 'pulses' of extra-terrestrial origin. Soon after the first pulsar was discovered, one was found at the centre of the Crab nebula, at the site of the Chinese supernova explosion. The Crab pulsar flashes on and off about 30 times a second, emitting energy across most of the electromagnetic spectrum. Pulsars were originally designated by the acronym LGM – standing for little green men – since they were first suspected to be signals from an extra-terrestrial civilisation. The truth is now believed to be somewhat less romantic – they are now thought to be rapidly rotating neutron stars!

It was Tommy Gold of Cornell, USA, who first realised that pulsars could be understood as rotating neutron stars. The rate of rotation required, however, was very much greater than that of normal stars. But, just as a skater draws in her arms to make a slow spin become a rapid spin (an elegant demonstration of angular momentum conservation) so too does the rotation of a star increase as it collapses to form a neutron star. Moreover, the magnetic fields of the star are also increased to a much higher value by this collapse. Now, as shown in Fig 10.8, the magnetic poles will not usually coincide with the poles of the axis of rotation. By a rather complicated mechanism involving both the magnetic and electric fields of the neutron star, it is believed that an intense narrow beam of radiation is produced in the direction of the magnetic axis. It is this beam of radiation sweeping across the Earth as the neutron star rotates that causes the observed 'pulsing' of the pulsar.

A neutron star is an amazingly compact and dense object. Nevertheless, the immense gravitational forces generated within such an object are countered by the neutron Pauli principle. However, it is believed that if the star is sufficiently massive (more than about three times the mass of our sun) even the Pauli principle for the quarks inside the neutrons cannot prevent the star collapsing to form an even more bizarre object, a 'black hole' (Fig 10.9). Such objects are permitted by Einstein's 'general theory of relativity' and correspond to a special type of solution to Einstein's equations. To form a black hole requires enormously high densities. For example, for our Sun to become a black hole, it would need to be compressed to the size of a ball about four miles in diameter. Once a star has been compressed smaller than a critical radius, the so-called 'Schwarzschild radius', the effects of gravity are so strong that nothing, not even light, can escape. It truly is a black hole!

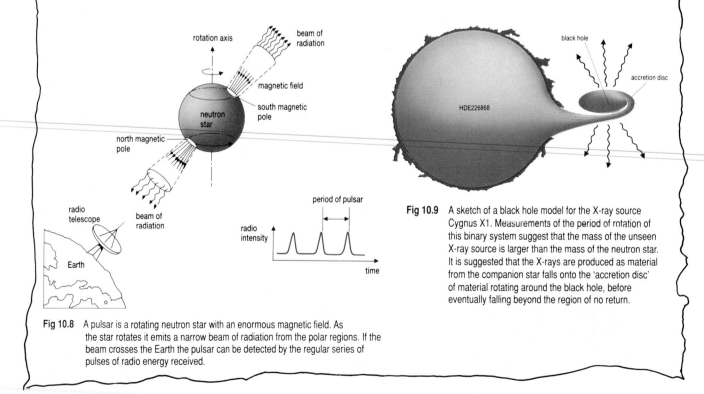

Fig 10.8 A pulsar is a rotating neutron star with an enormous magnetic field. As the star rotates it emits a narrow beam of radiation from the polar regions. If the beam crosses the Earth the pulsar can be detected by the regular series of pulses of radio energy received.

Fig 10.9 A sketch of a black hole model for the X-ray source Cygnus X1. Measurements of the period of rotation of this binary system suggest that the mass of the unseen X-ray source is larger than the mass of the neutron star. It is suggested that the X-rays are produced as material from the companion star falls onto the 'accretion disc' of material rotating around the black hole, before eventually falling beyond the region of no return.

10.5 A neutron star is an astronomical object, very massive, and consisting entirely of neutrons. Explain why, for such an object, we can ignore the surface and Coulomb terms in the binding energy formula.

10.6 Show that, for a neutron star, the symmetry term $d(N–Z)^2/A^2$ reduces simply to d.

10.7 Use values for m_n (the mass of a neutron) and r_0, together with equation (6) to find the smallest value of A for a neutron star. What is the mass of such a star? How does this compare with the mass of our Sun? (The Sun's mass = 2.0×10^{30} kg.)

10.8 **(a)** The radius of our Sun is about 7×10^8 m. Calculate its density.
   (b) If the Sun became a neutron star, estimate its radius and density.
   (c) If it became a neutron star, what would you estimate its density to be?

10.3 MATTER AND ANTIMATTER

Fig 10.10 Paul Dirac, the physicist whose theory of electrons led him to predict the existence of antimatter.

In Chapter 5 you found out something about antimatter. The particles known as positrons and antineutrinos are particles of antimatter. But what is the history of our knowledge of antimatter? And what is the history of antimatter itself in the Universe?

The first theoretical predictions of the existence and nature of antiparticles were made by an English physicist, Paul Dirac (Fig 10.10), in 1928. He combined the ideas of quantum theory and relativity in a theory of electrons which suggested that positively charged particles should also exist, with properties otherwise similar to those of the electron.

Any theory of antimatter must explain why it is not commonly found. Dirac's theory proposed that electrons could occupy both positive and negative energy levels; an electron in a negative energy level is a positron. Why do we not come across antimatter more frequently? Because all the negative levels are already filled. The electrons with negative energies cannot readily change their energy, and so we cannot easily observe these negative states. However, if an electron in a negative energy state is given sufficient energy, it can move into a positive state and become a particle, rather than an antiparticle.

These ideas were originally treated with scepticism. Dirac's ideas were extended to give a picture of matter and antimatter: pairs of particles having opposite charges, created out of pure energy. Where should one look for such unusual creatures?

The answer is, in high energy cosmic rays. In 1932 the first positrons were detected in cosmic ray showers. Later it was found that positrons and electrons annihilated one another in collisions. All that was left was a burst of gamma rays. Nowadays such events can be observed routinely in bubble chambers and other particle detectors. Fig 10.11 shows the creation of an electron-positron pair; a photon of energy has given rise to the two oppositely charged particles.

Is their much antimatter in the Universe? It is often suggested in science fiction that their might be antimatter galaxies, perhaps an anti-solar system with anti-planets identical to our own and anti-students studying anti-nuclear physics textbooks. The only distinction between this anti-Earth and our own Earth is that it would be a mirror image; reach out to shake hands with your anti-person image, and the two of you would disappear in a flash!

So why can we not answer such a simple question about the Universe? The problem with detecting antimatter is that we rely on light (and other forms of electromagnetic radiation) to tell us about the stars and galaxies

Fig 10.11 A bubble chamber photograph showing creation of an electron-positron pair.

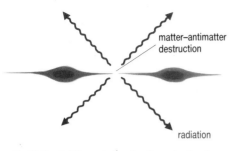

Fig 10.12 Colliding galaxies of matter and antimatter might be driven apart by the vast amounts of energy produced where their outer fringes meet.

which we observe. And we cannot tell whether light is emitted by matter or antimatter, because light is its own antiparticle. A photon is indistinguishable from an antiphoton.

Antimatter cannot exist for any length of time on the Earth. Similarly there cannot be anti-planets in our solar system. But might there be anti-galaxies? Current theories of the origin of the Universe suggest that, at the time of the Big Bang, both matter and antimatter came into existence. There was an imbalance between the two, with more matter formed than antimatter. It is possible that anti-galaxies still exist. Fig 10.12 shows one idea of what would happen if a galaxy collided with an anti-galaxy.

High energy particle experiments

Dirac's theory predicted the existence of positrons, and within a short time they were observed. This is an interesting example of the way in which theory often leads and guides experiments in physics. Nowadays many physicists are routinely engaged in experiments using high-energy particle accelerators, beams of particles and antiparticles, and so on. They are testing theories of the forces which act between particles, and theories of the ways in which matter itself came into existence.

Fig 10.13 The giant experiment in which protons and antiprotons collide at CERN.

Fig 10.14 The spray of particles and antiparticles produced when protons and antiprotons collide.

Fig 10.13 shows the proton – antiproton accelerator at the European CERN laboratory at Geneva. In Fig 10.14 you can see a collision pattern of particles from this accelerator – an experiment which confirmed one of the theories of the way in which particles interact. The particles and antiparticles being sought are very tightly bound, and increasingly high energies are needed to rip matter apart to reveal its fundamental structure.

Fig 10.15 shows a reaction produced by a beam of neutrinos. Considerable expertise is needed to interpret the resulting shower of particles.

Fig 10.15 A neutrino reaction seen in the Big European Bubble Chamber at CERN. A beam of neutrinos (invisible in the picture) entered from the left, and interacted with a quark to produce this dramatic spray of particles.

QUESTIONS

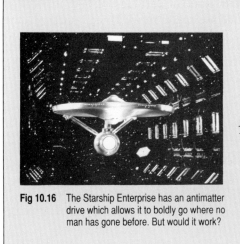

Fig 10.16 The Starship Enterprise has an antimatter drive which allows it to boldly go where no man has gone before. But would it work?

10.9 Large accelerators are very inefficient at producing antimatter; they require several megawatts of power when running and produce less than a nanogram of antimatter in a year. The Starship Enterprise (Fig 10.16) uses an antimatter drive to power it. Estimate the energy released in the destruction of a nanogram of antimatter. Estimate the speed of a large starship if its kinetic energy were derived from the destruction of a tonne of antimatter.

10.10 Think about the problems of containing antimatter, if it is to be used to power a starship. In Chapter 9 you looked at the problems associated with containing plasma for fusion reactions. Now design a containment system for an antimatter drive. Show how it can store antimatter and release controlled amounts of antimatter into a reaction chamber, where it must collide with a beam of matter. The resulting matter–antimatter annihilation must result in a release of energy to drive the starship.

10.4 PARTICLES AND FORCES, QUARKS AND GUTs

u	d
up	down
$Q = +\tfrac{2}{3}e$	$Q = -\tfrac{1}{3}e$
c	s
charm	strange
$Q = +\tfrac{2}{3}e$	$Q = -\tfrac{1}{3}e$
t	b
top	bottom
$Q = +\tfrac{2}{3}e$	$Q = -\tfrac{1}{3}e$

Fig 10.17 The six members of the quark family.

e	ν_e
electron	electron-neutrino
$Q = -e$	$Q = 0$
μ	ν
muon	muon-neutrino
$Q = -e$	$Q = 0$
τ	ν_τ
tau	tau-neutrino
$Q = -e$	$Q = 0$

Fig 10.18 The six members of the lepton family. For each particle, there is also an antiparticle.

In the first nine chapters of this book we have built up a picture of matter as consisting of combinations of particles; in particular, we can think of the atom as being built up from electrons, protons and neutrons. These particles now seem very familiar and most young people study a simple model of the atom before they are 16.

However, this picture represents a simplified version of what was well-established by 1920. Since then many experiments, including those using high-energy particle accelerators, have revealed that protons and neutrons are themselves divisible. They are made up of even more fundamental particles called **quarks**.

In Fig 10.17, you can see the members of the quark family. Perhaps the most surprising things about quarks are their names – up, down, charm, strange and so on. These are something of a physicists' joke. They refer to properties of particles made from these different quarks. Since we cannot really describe the curious properties of such particles, physicists say they have different flavours, such as charm and strangeness. It is their way of making sense of something which is truly inexpressible.

Electrons are very different from protons and neutrons; they appear to be truly indivisible. They are point-sized particles – they have no size. They belong to a family of particles called **leptons**, which means 'light ones', since they have very little mass. Neutrinos also belong to this family (see Fig 10.18).

g	γ	W	Z	g
gluon	photon	weak	bosons	gravitino
$Q = 0$	$Q = 0$	$Q = \pm e$	$Q = 0$	$Q = 0$

Fig 10.19 The five gauge bosons – the particles which transmit the fundamental forces of nature.

A final group of particles are the **gauge bosons** (Fig 10.19). These are the particles which correspond to the forces which bind matter together: gluons for the strong nuclear force, photons for the electromagnetic force, and so on.

NUCLEAR PHYSICS TODAY

Building a proton

Protons and neutrons are the most familiar particles of a group called **baryons** – 'heavy ones'. All baryons are made up of three quarks. In the case of protons and neutrons, the only quarks involved are the up and down quarks.

Using the information about quark charges given in Fig 10.17 you can work out what combination of quarks is needed for each of these baryons.

QUESTIONS

10.11 What combinations of up and down quarks are required to make a proton and a neutron?

10.12 When a neutron decays to a proton, an electron and an antineutrino are emitted. Can you explain what change of quarks must occur within the neutron for this to happen?

Fig 10.20 James Clerk Maxwell, the Scottish physicist whose theory of electromagnetism united the electrical and magnetic forces.

Unifying the forces of nature

We are used to talking about 'electromagnetism' as a single phenomenon. Electricity and magnetism are, we know, closely related. An electric current disturbs a magnetic compass; a changing magnetic field induces a current in a conductor. However, this unity of electricity and magnetism is not immediately obvious. A young child might play with magnets, and with batteries and bulbs, without ever suspecting that the two were related.

It was Oersted in 1820 who showed, experimentally, the connection. Forty years later Maxwell (Fig 10.20) developed his mathematical theory which was capable of describing both electricity and magnetism, and which showed that light, too, was part of the same phenomenon.

This illustrates one of the driving passions of physicists; to simplify a complex web of observations into a single, all-embracing theory. Since the time of Maxwell great strides have been made in bringing together our ideas of the other forces of nature. The 1960s saw the unification of the electromagnetic and weak forces within a single theory, by Abdus Salam and Steven Weinberg (Fig 10.21). Their theory predicted the existence of the W and Z bosons, and these were subsequently observed in proton–antiproton collision experiments. This is another example of theory leading and guiding experimentation in Physics.

Fig 10.21 Abdus Salam, whose theory unified electromagnetism and the weak nuclear force. He was the first Pakistani and the first Muslim to win a Nobel Prize.

Fig 10.22 shows the progression in the development of theories of the forces of nature. Grand Unified Theories, or GUTs, have shown how the strong force can be incorporated in the picture. Now what is needed is a further development to bring the last fundamental force, gravity, into the picture.

Our understanding of the particles and forces of nature have changed dramatically during the last 100 years. A simple picture of indivisible atoms gave way to the nuclear picture of the atom. New forces were discovered to explain this picture. Experiments showed the existence of many more fundamental particles; theories were developed to simplify this to a picture of leptons, quarks and bosons. Seemingly disparate forces have been brought together in difficult theoretical frameworks.

Such theories are regarded by physicists as simplifications of complex observations of the natural world. They are among the greatest of human intellectual achievements.

Fig 10.22 Progress in the theoretical unification of the fundamental forces of nature.

EXAMINATION QUESTIONS: Theme 3

T3.1

State the identity of α, β^+, β^- and γ radiations.

The unstable isotope $^{14}_{6}C$ undergoes β^- decay with a half-life of 5730 years (= 3.01×10^9 minutes) to a nucleus $^A_Z X$.

(a) State the values of Z and A, and name another particle given out in the decay.

(b) Describe the measurements you would make to verify that the $^{14}_{6}C$ isotope emits β^- radiation, rather than α, β^+ or γ radiations.

(c) Natural carbon (mainly ^{12}C) in a living organism contains a fraction 1.3×10^{-12} of ^{14}C atoms. Calculate the number of ^{14}C decays per minute you would expect in 1.0 g of carbon from a living organism.

(d) A fossil sample containing 2.0 g of carbon gives a measured decay rate of 7.5 decays per minute. Estimate the age of the fossil, stating your assumptions. Give one reason why the result will be only an estimate of the true age of the fossil.

(e) Suggest measurements from which the half-life of 5730 years for $^{14}_{6}C$ could be determined.

(O&C 1987)

T3.2

(a) The path of a charged particle through a cloud chamber appears as a thin white line. What does this line consist of and how is it formed?
   How may the tracks of beta particles be distinguished from those of alpha particles in the absence of any deflecting electric or magnetic fields?

(b) The diagram shows an exaggerated view of the track of a beta particle as it passes through a thin sheet of lead in a cloud chamber. A uniform magnetic field acts *into* the paper, through the whole region shown in the diagram, and the beta particle moves in the plane of the paper.

(i) Is the beta particle travelling in the direction P to Q or Q to P? Explain your reasoning.

(ii) Does this beta particle carry a positive or negative charge? Explain your reasoning.

(c) A $_{19}^{40}$K atom has mass 39.964 001 u and a $_{20}^{40}$Ca atom has mass 39.962 582 u where 1 u equals $1.660\,4 \times 10^{-27}$ kg. The speed of light in vacuum is 3.00×10^8 m s^{-1}.

   (i) Write down an equation which describes the decay of a $_{19}^{40}$K atom into a $_{20}^{40}$Ca atom and identify the particle emitted by the nucleus.

   (ii) Why, in the decay process, does the number of extranuclear electrons increase by one?

   (iii) Calculate the energy released from the $_{19}^{40}$K nucleus during the decay.

   (iv) Describe one possible mode of decay of a $_{19}^{40}$K nucleus into a $_{18}^{40}$Ar nucleus.

(ULSEB 1986)

T3.3
Answer the following, which all relate to nuclear fission.

(a) Use the binding energy equation of the liquid drop model to explain the source of energy release in the fission of a nucleus of uranium.

(b) Why are the daughter products of fission normally of different mass numbers?

(c) Why does a nucleus of ^{235}U readily undergo fission after absorption of a thermal neutron?

(d) Why does almost any substance become radioactive if placed inside a fission reactor, even if enclosed in a sealed container?

(e) State and explain what form of radioactivity you would expect to be produced by the process in question (d) above.

(JMB 1980)

T3.4
(a) Explain the action of a moderator in a thermal neutron fission reactor. Give **two** physical factors which are important in making a choice of moderator material.

(b) Discuss why cadmium is commonly used to make control rods in such a reactor. Explain how control rods are used.

(c) Spent fuel elements from a nuclear reactor contain appreciable concentrations of many radioactive isotopes. Explain why this is so.

(JMB 1979)

T3.5
(a) Explain how a controlled, self-sustaining chain reaction is maintained in a nuclear reactor which uses natural uranium as a fuel.

(b) If the total decay rate of the radioactive products in a nuclear reactor is 10 times the fission rate, estimate the steady state activity in disintegrations per second in a reactor working at 30 per cent efficiency and delivering 1.0 GW. It may be assumed that 2.0×10^8 eV of energy is released in each fission process. (1 eV = 1.6×10^{-19} J.)

(JMB 1979)

T3.6

(a) Write notes on the safety aspects of the use of nuclear reactors for power production under the following headings. (i) Normal operation. (ii) Emergency shutdown. (iii) Waste disposal, including the treatment of spent fuel rods.

(b) There is evidence that samples of uranium mined in Gabon in Africa have acted as fuel in a natural nuclear reactor. State one difference you would expect between purified samples of this uranium and purified samples of uranium mined elsewhere. Explain your reasoning.

(JMB 1985)

T3.7

It is possible in principle to obtain energy from the fusion of light nuclei. One suitable reaction is

$$^3H + {}^2H \rightarrow X + Y + Q$$

where X and Y represent the two products of the reaction and Q the energy release. Where necessary using data from the table below, answer the following questions.

(a) Identify the products X and Y.

(b) Calculate the value of Q in MeV

(c) Calculate the kinetic energy of each of the product nuclei, X and Y, assuming that the initial nuclei are at rest.

(d) Give two physical reasons why it has so far proved impossible to build a practical power station using fusion as an energy source.

In the following table, which lists a selection of nuclides, the mass of an isotope is given for a neutral atom of the substance and is quoted in unified atomic mass units, u. (1 u is equivalent to 931.3 MeV.)

Name	Symbol	Charge number Z	Mass number A	Atomic mass M
electron	$e^{\pm}$	± 1	–	0.000 55
proton	p	1	1	1.007 28
neutron	n	0	1	1.008 67
hydrogen	^1H	1	1	1.007 83
	^2H	1	2	2.014 10
	^3H	1	3	3.016 05
helium	^3He	2	3	3.016 03
	^4He	2	4	4.002 60
lithium	^6Li	3	6	6.015 13
	^7Li	3	7	7.016 01

(JMB 1986)

Appendix A

BOOKS AND OTHER RESOURCES

1. **Standard textbooks**
    Most standard A-level textbooks include an introduction to nuclear physics. In particular, they contain the basic ideas of radioactivity, with which you should be familiar. Two such books are:

    Physics, by R Hutchings. Macmillan, 1989.
    Physics, by Tom Duncan. John Murray, 1982.

2. **Other textbooks**
    Other books written at a similar level to this include:

    Modern physics, by D E Caro, J A McDonell and B M Spicer. Arnold, 1978.
    Nuclear physics, by Joyce Dacre. Heinemann, 1989.

3. **More advanced textbooks**
    Some undergraduate textbooks may provide useful source material, although most are written at an inappropriate level. One relatively approachable text is:

    Nuclear physics, by Irving Kaplan. Addison Wesley, 1963.

4. **Fundamental forces, fundamental particles**
    There are several popular accounts of the ideas touched upon in Chapter 10. Amongst these are:

    The accidental universe, by P C W Davies. Cambridge University Press, 1982.
    The cosmic onion, by F Close. Heinemann, 1983.
    The forces of nature, by P C W Davies. Cambridge University Press, 1986.
    Physics: Particles, imaging and nuclei, ed. John Harris, Nuffield Advanced Physics. Longman, 1986.
    The quantum universe, by Tony Hey and Patrick Walters. Cambridge University Press, 1987.

5. **Nuclear power, nuclear weapons**
    In addition to the leaflets, booklets and books produced by pro-nuclear and anti-nuclear groups, there are many popular accounts of the nuclear industry and its products. These include:

    Brighter than a thousand suns, by Robert Jungk. Penguin, 1964.
    Hiroshima, by John Hersey. Penguin, 1964.
    Power production: What are the risks? by J H Fremlin. Oxford University Press, 1987.
    The nuclear issue, by J J Wellington. Blackwell, 1986.
    Nuclear power, by Walter Patterson. Pelican, 1976.
    Nuclear power for beginners, by Stephen Croall and Kaianders Sempler. Writers and Readers, 1985.
    Radiation risks: an evaluation, by David Sumner. Tarragon Press, 1987.
    Understanding nuclear power, by H A Cole. Gower Technical Press, 1988.

6. **Data books**
   Book of data, Revised Nuffield Advanced Science. Longman, 1984.
   Science data book, by R M Tennent. Oliver and Boyd, 1971.

7. **SATIS 16–19**
   Several units produced by the SATIS (Science and Technology in Society) 16–19 Project relate to aspects of nuclear technology. These are published by the Association for Science Education.

8. **Posters**
   In addition to those produced by the UKAEA (see their catalogue), two useful posters are produced by the *New Scientist*:

   The elementary particles
   Radioactivity

   The most detailed chart of the nuclides, published by ESSO, is not currently in print.

9. **Videos**
   The UKAEA provides a series of videos relating to all aspects of the nuclear power industry. These include:

   Isotopes in action
   The fast reactor
   The fast reactor fuel cycle
   The management of nuclear waste
   The nuclear fuel cycle
   Plutonium and the fast reactor
   Principles of fission

   A programme from the Granada TV series *Experiment* shows alpha particle scattering by metal foils, and provides the opportunity to analyse real data:

   Rutherford scattering

   Two videos from Friends of the Earth deal with nuclear power in the context of renewable energy sources:

   Energy without end
   Positive charge

10. **Software**
    Two pieces of computer software, produced by the UKAEA, are referred to in the text (see Appendix B for address):

    Nuclides data base
    Nuclear reactor simulation

    A similar reactor program produced by the BBC, is:

    Inside science: nuclear power

11. **Places to visit**
    There is a large public exhibition at the Visitors' Centre at Sellafield. In addition, the UKAEA will provide facilities for school and college visits to its establishments.
       There is a nuclear physics/nuclear energy gallery at the Science Museum in London. The exhibits include many items of historical interest; others reflect the considerable funding for the gallery provided by the nuclear industry.

Appendix B

USEFUL ADDRESSES

1. **Nuclear power**

   Several organisations produce materials useful for providing background information about nuclear power. Much of this is partisan.

   British Nuclear Fuels plc, Risley, Warrington WA3 6AS.

   Central Electricity Generating Board, Sudbury House, 15 Newgate Street, London EC1A 7AU.

   Centre for Alternative Technology, Machynlleth, N. Wales.

   Department of Energy – Information Branch, Thames House South, Millbank, London SW1P 4QJ.

   Ecoropa, Crickhowell, Powys, Wales.

   Friends of the Earth Trust Ltd, 26-28 Underwood Street, London N1 7JQ.

   Greenpeace, 30-31 Islington Green, London N1 8BR.

   Nuclear Electricity Information Group, 23 Buckingham Gate, London SW1E 2LB.

   UKAEA Information Service, 11 Charles II Street, London SW1Y 4QP.

   UK Nirex Ltd, Curie Avenue, Harwell, Didcot, Oxfordshire OX11 0RH.

2. **Nuclear weapons**

   Organisations concerned with nuclear weapons produce materials which are concerned more with nuclear strategy and politics than with nuclear technology.

   Bradford University School of Peace Studies, Richmond Road, Bradford BD7 1DP.

   Campaign for Nuclear Disarmament, 22–24 Underwood Street, London N1 7JQ.

   Scientists Against Nuclear Arms, 9 Poland Street, London W1V 3DG.

   Coalition for Peace through Security, 27–31 Whitehall, London SW1A 2BX.

   Defence Secretariat Ministry of Defence, Whitehall, London SW1.

Appendix C

USEFUL CONSTANTS AND CONVERSION FACTORS

quantity	symbol	value
speed of light in a vacuum	c	3×10^{8} m s^{-1}
permittivity of free space	ε_0	8.85×10^{-12} F m^{-1}
Planck constant	h	6.63×10^{-34} J s
Gravitational constant	G	6.673×10^{-11} N m^2
Avogadro constant	N_A	6.022×10^{23} mol^{-1}
electron charge	e	1.60×10^{-19} C
electron mass	m_e	9.110×10^{-31} kg
proton mass	m_p	1.673×10^{-27} kg
neutron mass	m_n	1.675×10^{-27} kg

Conversion factors

1 eV = 1.602×10^{-19} J

1 u = 1.661×10^{-27} kg

1 u = 931.3 MeV

ANSWERS TO QUESTIONS AND ASSIGNMENTS

Chapter 1

1.1 6.70×10^{-19} J

1.2 1 keV; 1.6×10^{-16} J; 1.87×10^7 m s^{-1}

1.3 1.75 eV

1.4 θ decreases as p increases

1.5 θ decreases as v increases

1.6 8×10^{-13} J

1.7 5×10^{-14} m

1.8 Radius $< d$

1.9 In order that the scattering angle θ can be determined accurately, the incident beam must be very finely collimated

1.10 More energetic particles would be deflected less as they pass the nucleus, so fewer would be back-scattered

1.11 Gamma rays

1.12 Increase it

1.13 Increase it

1.14 Increase cathode–anode pd

1.15 2.0×10^{-10} J; 1240 MeV

1.16 1240 MV

1.17 3.0×10^{-15} m

1.18 4.6×10^{-15} m

1.19 5.3×10^{-15} m

1.20 Since neutrons do not experience the Coulomb force, they can penetrate closer to the nucleus

1.21 V increases uniformly with A

1.22 r on y-axis, $A^{1/3}$ on x-axis, gradient = r_0, intercept = 0

1.23 1.3×10^{-15} m

1.24 Approx 10^{17} kg m^{-3}; 10^{14} times

1.25 1 u = 1/Avogadro's number (in g)

1.26 3.90407×10^{-25} kg

1.27 2.3259×10^{-26} kg; 14.00324 u

Chapter 2

2.1 230 N

2.2 1.37×10^{29} m s^{-2}

2.3 1.87×10^{-34} N

2.4 $F_e : F_g = 1.7 \times 10^{36}$

2.5 The region of the graph where F is negative is where the force is attractive

2.6 30 000 N

2.7 $r \sim 0.6$ fm; Coulomb $F = 640$ N

2.8 A negative force is repulsive

2.9 1.5×10^{-11} N

2.10 $r \sim 0.4$ nm; Coulomb $F \sim 1.4 \times 10^{-9}$ N (note that this is considerably greater than the van der Waals force at this separation)

2.11 1.5×10^{-11} J

2.13 The flat bottom indicates zero force on a proton within the nucleus; vertical walls imply infinite force; sloping walls imply finite force, which is more realistic

2.14 10 MeV

Chapter 3

3.1 $^{35}_{17}Cl$ (75.5%); $^{37}_{17}Cl$ (24.5%)

3.2 5570 years, β

3.3 3

3.4 1.17 MeV; 1.33 MeV

3.5 $^{209}_{83}Bi$

3.6 $Z = 20$ approx

3.7 $N : Z = 1.5$ approx

3.8 The neutrons help to separate the protons, and thereby reduce their mutual electrostatic repulsion, which would otherwise make the nucleus unstable

3.9 Isotopes: in line, vertically. Isotones: in line, horizontally

3.10 Even Z, even N

3.11 $^{2}_{1}H$, $^{6}_{3}Li$, $^{10}_{5}B$, $^{14}_{7}N$

3.12 $^{4}_{2}He$

3.13 Above the line of stability

3.14 β⁻: above the line of stability.   β⁺: below the line.

3.15 Z decreases by one, N increases by one

3.16 $^{208}_{83}Bi$

3.17 See Fig 3.4

3.18 A is $4 \times$ integer

3.19 4n+2

3.21 α, $^{231}_{90}Th$, 7.1×10^{8} years

3.22 $A = 207$

3.23 7 α decays, 4 β decays

3.24 The 4n+3 series must have existed, but has now decayed, since the age of the Universe is many times the half-life

Chapter 4

4.1 15.990 523 u

4.2 11.996 706 u; 2.75×10^{-4} (1 part in 3640)

4.3 19.992 440 u

4.4 0.007 187 u

4.5 0.132 613 u

4.6 $^{16}_{8}U$

4.7 9×10^{16} J

4.8 1.1×10^{-17} kg

4.9 1.5×10^{-10} J;   931 MeV

4.10 **(a)** 2500 J, 2.8×10^{-14} J,   approx 5 parts in 10^{15}

(b) 5×10^{5} J, 5.6×10^{-12} kg,   approx 1 part in 10^{13}

(c) 1 MJ, 1.1×10^{-11} kg,   approx 2 parts in 10^{13}

4.11 **(a)** 28.29 MeV **(b)** 7.07 MeV/nucleon

4.12 2.23 MeV/nucleon;   7.72 MeV/nucleon

4.13 $^{16}_{8}U$

4.14 21.60 MeV

4.15 The volume term

4.16 The surface term. In heavy nuclei, a smaller fraction of nucleons are near the surface

4.17 The Coulomb term is greater for large A, since more protons are packed closely together

4.18 See Table 4.2 opposite.

The surface term is most significant (proportionately) for the lightest nuclide, $^{7}_{3}Li$. The Coulomb and symmetry terms are most significant for the heaviest nuclide, $^{209}_{83}Bi$. $^{56}_{26}Fe$ is the most stable nuclide.

Table 4.2

Nuclide	Z	N	A	Volume term/MeV	Surface term/MeV	Coulomb term/MeV	Symmetry term/MeV	ΔE MeV	$\Delta E?A$ MeV/nucleon
$^{7}_{3}\text{Li}$	3	4	7	98	–47.6	–2.7	–2.8	44.9	6.4
$^{56}_{26}\text{Fe}$	26	30	56	784	–190	–102	–6	486	8.7
$^{209}_{83}\text{Bi}$	83	126	209	2926	–458	–673	–171	1624	7.8

4.19 $^{15}_{6}\text{C}$: $\Delta M = 0.114\,335$ u; $\Delta E = 106.5$ MeV; β^- emission
$^{15}_{7}\text{N}$: $\Delta M = 0.123\,987$ u; $\Delta E = 115.5$ MeV; most tightly bound
$^{15}_{8}\text{O}$: $\Delta M = 0.120\,183$ u; $\Delta E = 111.9$ MeV; β^+ emission

4.20 $^{16}_{8}\text{O}$: $\Delta E/A = 7.97$ MeV/nucleon; most abundant
$^{17}_{8}\text{O}$: $\Delta E/A = 7.75$ MeV/nucleon
$^{18}_{8}\text{O}$: $\Delta E/A = 7.77$ MeV/nucleon

Chapter 5

5.1 8.17 MeV
5.2 8.0 MeV; kinetic energy of recoil of daughter nucleus
5.4 $E_{k\alpha} = 6.76$ MeV; $E_{kY} = 0.113$ MeV
5.6 0.783 MeV
5.7 10.8 minutes
5.8 No. Free protons have a half-life of 10^{30} years – they are very stable
5.9 Proton-rich nuclei
5.10 $^{1}_{1}\text{p}_0 + ^{0}_{-1}\text{e}_0 + \nu + Q$
5.12 $Q < 0$
5.13 5.52 MeV
5.14 β^- emission: 0.57 MeV
β^+ emission: 0.65 MeV
Electron capture: 1.68 MeV
5.15 No; it decays by electron capture
5.16 $\Delta E/A$ increases

Chapter 6

6.1 All about the same
6.2 Two different energies of α particles are present
6.3 $E_{k\alpha}$ increases uniformly with $t_{1/2}$
6.4 A straight line graph
6.5 6.0 MeV
6.6 1.6×10^7 m s^{-1}
6.7 15.9 fm
6.8 6×10^{28} times
6.9 $^{101}_{44}\text{Ru}$
6.10 $Z < 44$
6.11 $Z > 44$
6.12 The neutrino carries off 99.999% or so of the energy released
6.13 X-rays, characteristic of the daughter atom
6.14 4.59 MeV
6.15 2.60 MeV
6.16 1.01 MeV, 0.83 MeV, 0.18 MeV; $E_1 = E_2 + E_3$
6.17 1.77 MeV
6.18 1.17 MeV, 1.33 MeV; 0.32 MeV; 2.82 MeV

Chapter 7

7.1 Lower
7.4 11 mm
7.5 1/1 000 000
7.7 1.6×10^{-15} A
7.8 625 particles per second

Chapter 8

8.1 92 and 141
8.2 $\frac{1}{10000}$
8.4 3 neutrons
8.5 No
8.6 173.2 MeV
8.15 0.52 MeV
8.16 2200 m s^{-1}
8.22 **(a)** 3.6×10^{30} eV; **(b)** 5.7×10^{11} J; **(c)** 159 000 kW
8.23 1.55×10^{10} nuclei
8.24 41.6%
8.27 173 MeV
8.34 0.43%
8.35 3.1%
8.36 A few tonnes

Chapter 9

9.1 **(a)** 5.5 MeV **(b)** 17.3 MeV **(c)** 0.13 MeV
9.2 β^+, ν
9.3 6.26 MeV; 0.53 MeV
9.4 1 MeV : 3 MeV
9.5 **(a)** 2.3 N **(b)** 144 keV
9.6 1.1×10^9 K
9.8 **(a)** 1.2×10^{56} **(b)** 10^{10} years **(c)** 5000 million years

Chapter 10

10.6 $Z = 0$, so $A = N$
10.7 3.1×10^{55}; 5×10^{28} kg; 1/40 solar mass
10.8 **(a)** 1400 kg m^{-3}
(b) 13.8 km; 5.7×10^{17} kg m^{-3}
(c) 5.8×10^{19} kg m^{-3}
10.11 proton = uud, neutron = udd
10.12 One u becomes a d

Index